Marina Ismael Michen

Co-Extrusion of Piezoelectric Ceramic Fibres

Schriftenreihe des Instituts für Keramik im Maschinenbau
IKM 57

Institut für Keramik im Maschinenbau
Karlsruher Institut für Technologie

Co-Extrusion of Piezoelectric Ceramic Fibres

von
Marina Ismael Michen

Institut für Keramik im Maschinenbau (IKM),
Karlsruher Institut für Technologie

Dissertation, Karlsruher Institut für Technologie
Fakultät für Maschinenbau,
Tag der mündlichen Prüfung: 26.11.2010

Impressum

Karlsruher Institut für Technologie (KIT)
KIT Scientific Publishing
Straße am Forum 2
D-76131 Karlsruhe
www.ksp.kit.edu

KIT – Universität des Landes Baden-Württemberg und nationales
Forschungszentrum in der Helmholtz-Gemeinschaft

KIT Scientific Publishing 2011
Print on Demand

ISSN: 1436-3488
ISBN: 978-3-86644-653-3

Co-Extrusion of Piezoelectric Ceramic Fibres

Zur Erlangung des akademischen Grades eines

Doktors der Ingenieurwissenschaften

der Fakultät für Maschinenbau
Karlsruher Institut für Technologie (KIT)
eingereichte

genehmigte
Dissertation

von

Marina Ismael Michen
aus Ribeirão Preto / Brasilien

Tag der mündlichen Prüfung: 26.11.2010
Hauptreferent: Professor Dr. rer. nat. M. J. Hoffmann
Korreferent: Professor Dr. rer. nat. T. J. Graule

To the memory of my grandpa

ACKNOWLEDGEMENTS

This research was developed in collaboration between the Laboratory for High Performance Ceramics (Group of Dr. Frank Clemens) in the Swiss Federal Laboratories for Materials Testing and Research (EMPA, Switzerland) and the Institute of Ceramics in Mechanical Engineering in the Karlsruhe Institute of Technology (KIT, Germany). It was supported by internal funding of the EMPA, under the project number 880109.

As with every work of science, there are many people who have helped to develop this project. I am especially indebted to Prof. Dr. rer. nat. M. J. Hoffmann and Prof. Dr. rer. nat. T. J. Graule for the opportunity provided. Additionally, I appreciate their critical questions, important advices and valuable suggestions during this work.

I am exceptionally grateful to Dr. Frank Clemens for his assistance in developing and producing this work. Thank you for the patience, for the (long) discussions, confidence and advices.

In particular, I express my gratitude to Benjamin Michen for showing me the way to the library, for the patience, for the kindness shown in reading and constructively criticizing various parts of this work and, finally, for translating the abstract to German.

Furthermore, I would like to acknowledge the very interesting and scientific discussions which I had in the course of this project with Dr. Hans Kungl.

I am thankful to the master students that carried out their work under my co-supervision: Wilben M. Bohac, Viviane L. Bueno, Aneta Adamska and Robert Dittmer. I have also learned by them!

I express my sincere "Vielen Dank!" for Hans Jürgen Schindler for his support in the lab. Additionally, I am grateful to those who have taught/helped me in performing some measurements, especially: Dr. Juliane Heiber ("Grazie, Bella!"), Francisco Alvarez ("Gracias, Paco!"), Noemie van Garderen, ("Merci, Noe!") Beatrice Fischer, Peter Wyss and Dr. Daniela Suppiger.

Last, but certainly not least, the continual encouragement and support of my parents, sister, and friends is sincerely appreciated: "muito obrigada!"

Marina

ABSTRACT

The co-extrusion process involves the simultaneous extrusion of multiple materials, such as ceramic-thermoplastic compounds that can be used to manufacture fine-scaled piezocomposites, for both structural and functional applications. The principle of this technique consists of fabricating a macro-scaled preform composite and afterwards extruding it through a die to generate fine structures which are identical in their geometry and composition to the original preform, albeit with reduced cross sectional dimensions. The process can be repeated to reduce the size and multiply the number of shaped patterns. The main challenge however, is the adjustment of the flow mechanics of the materials to be co-extruded, towards defect-free structures.

The present work successfully developed a methodology for fabricating lead zirconate titanate [PZT] thin solid- and hollow-fibres by the thermoplastic co-extrusion process. The whole process chain, that includes: a) compounding, involving the mixing of ceramic powder with a thermoplastic binder, b) rheological characterizations, c) preform composite fabrication followed by co-extrusion, d) debinding and, finally, e) sintering of the body to near full density, is systematically described. The preform composite is composed by two different extrudable mixtures (feedstocks). The primary feedstock contains the ferroelectric ceramic which is to be micro-fabricated. The secondary contains a fugitive substance with the function to fill the space between the green filaments, provided that it can be completely removed after the micro-fabrication. A feedstock containing 58 vol. % of PZT was selected as the primary material, whereas two different fugitive materials have been used: carbon black (CB) and microcrystalline cellulose (MCC). Well preserved fibre morphologies with defined interfaces between the co-extruded materials have been achieved using the co-extrusion process when the rheological behaviour of the feedstocks and the extrusion parameters were correlated. In order to maintain the micro-fabricated fibres after debinding of the organic materials, it is important that the onset temperature for decomposition of the outer layer is lower than for the inner one. Thus, the generation of stresses due to the evolution of gaseous species is avoided. In the case of solid-fibres these stresses caused fractures in the green fibres, while hollow-fibres showed no damage.

To establish the developed co-extrusion process, the microstructure and the electromechanical response of the sintered co-extruded PZT fibres have been compared with PZT fibres of similar diameter (~ 250 µm) obtained by conventional extrusion. The presence of unfavourable impurities in the CB material led to compositionally modified-PZT fibres, inducing a drop of 40% in the electromechanical performance. However, the use of MCC as an alternative to the CB reached comparable microstructure and electromechanical properties to fibres processed by extrusion. Despite that the current research worked with PZT as the primary material, the method may be transferred to different ceramic materials to manufacture thin fibres and composites.

ZUSAMMENFASSUNG

Das Verfahren der Co-Extrusion beinhaltet die simultane Extrusion mehrerer Materialien wie zum Beispiel Keramik-Thermoplast Komposite, die zur Herstellung von feinen Piezoverbundwerkstoffe für strukturelle und funktionelle Anwendungen verwendet werden. Das Prinzip dieser Technik basiert auf der Herstellung einer sogenannten Preform, welche als vergrößertes Abbild des zu extrudierenden Bauteils definiert ist. Durch die Extrusion der Preform wird der Querschnitt dieser verkleinert wobei die Struktur erhalten bleibt. Der Vorteil der Preform-Methode liegt vor allem in der Herstellung von feinen Strukturen. Dieser Prozess kann wiederholt angewendet werden um den Querschnitt weiter zu reduzieren und gegebenenfalls die Anzahl der Einzelstrukturen im Verbundwerkstoff zu vervielfachen. Die Herausforderung liegt darin das Fließverhalten der beteiligten Materialien so einzustellen, dass die Struktur der Preform im Produkt erhalten bleibt.

Die hier vorgelegte Arbeit entwickelt erfolgreich die Methoden zur Herstellung von feinen Blei-Zirkonat-Titanat (PZT) Fasern und Hohlfasern auf Basis der thermoplastischen Co-Extrusionsmethode. Der gesamte Prozessablauf wird hier systematisch beschrieben, beginnend mit der Materialauswahl, über die Herstellung von Co-Extrusionsmassen sowie ihrer rheologischen Charakterisierung, bis hin zur Entbindung der Grünkörper und deren Sintern. Hervorragend erhaltene Fasern, mit sehr definierten Grenzflächen zwischen den co-extrudierten Materialien, konnten hergestellt werden so fern die rheologischen Eigenschaften der einzelnen Materialien aufeinander abgestimmt wurden. Um die Fasern während des Entbindungsprozesses des organischen Materials (wie z.B. Polymere oder Kohle) nicht zu zerstören, ist darauf zu achten, dass der Abbau der äußeren Schicht bei geringeren Temperaturen einsetzt als der der inneren Schicht. Dadurch kann die Entstehung von Spannungen, hervorgerufen durch austretende Gase, vermieden werden. Diese Gase waren verantwortlich für die Entstehung von

Rissen in Vollfasern, Hohlfasern hingegen zeigten sich unbeeinflusst.

Abschließend werden PZT-Fasern, hergestellt durch den entwickelten Co-Extrusionprozess, mit konventionell hergestellten PZT-Fasern gleichen Durchmessers (~ 250 µm) verglichen. Dabei werden die elektromechanischen Eigenschaften sowie das Gefüge untersucht. Die Anwesenheit von Verunreinigungen in der Kohle, welche als Füllstoff in der Preform eingesetzt wird, führt zu chemischen Veränderung des PZTs und verringert somit die elektromechanische Leistung um 40%. Die Verwendung von mikrokristalliner Cellulose als Füllstoff ermöglichte die Co-Extrusion von PZT-Fasern mit vergleichbaren Eigenschaften wie sie durch die konventionelle Extrusion erzielt werden. Obwohl diese Arbeit sich auf PZT-Fasern konzentriert, kann die hier entwickelte Methode auf weitere keramische Materialien angewendet werden um Fasern oder Verbundwerkstoffe herzustellen.

GLOSSARY OF TERMS

AFC	active fibre composite
BET	Brunauer-Emmett-Teller
CB	carbon black
DMA	dynamic mechanical analyzer
DSC	differential scanning calorimetry
EEA	poly(ethylene-co-ethyl acrylate)
Ext. V	extruded volume
LDPE	low density polyethylene
MCC	microcrystalline cellulose
MFI	melt flow index
MPB	morphotropic phase boundary
PiBMA	poly(isobutyl methacrylate)
PS	polystyrene
PSD	particle size distribution
PZT	lead zirconate titanate [$Pb(Zr_x, Ti_{1-x})O_3$]
PZZ	$PbZrO_3 + ZrO_2$
Ref.	reference
rpm	rotation per minute
SEM	scanning electron microscopy
SI	international system of units
SSA	specific surface area
St. acid	stearic acid
TGA	thermo-gravimetric analysis
vdW	van der Waals
YSZ	yttria-stabilized zirconia
XRD	X-ray diffraction
A	area
Al	allocation
c	sound velocity

$C(n)$	rheological mixing constant (torque-rheometer)
E	electric field
E_c	coercive field
E_f	energy of activation for viscous flow
f	frequency
F	force
k	instrument constant (concentric-cylinder and torque-rheometer)
$\overline{K}$	consistency index of the material
K_{eff}	effective electromechanical coefficient
L	length
M	torque moment
M_w	molecular weight
n	power law index
N	rotor speed (torque-rheometer)
N_A	Avogadro number
P	polarisation
P_r	remnant polarisation
P_s	spontaneous polarisation
P_{sat}	saturation polarisation
P_o	porosity
Q	volumetric flow rate
r	rotor radius (torque-rheometer)
R	reduction ratio
$\overline{R}$	universal gas constant
R_d	die radius (capillary-rheometer)
R_e	external radius (concentric-cylinder and torque-rheometer)
R_i	internal radius (concentric-cylinder and torque-rheometer)
S	strain
S_{max}	maximum strain

x

S_{rem}	remnant strain
S_{calc}	calculated shrinkage
S_{theo}	theoretical shrinkage
S_w	die swell
T	temperature
T_c	Curie temperature
T_g	glass transition temperature
T_m	crystalline melting temperature
T_{50}	temperature for 50% weight loss (TGA)
T_{max}	temperature corresponding to the maximum decomposition rate (TGA)
T_{onset}	temperature at which weight loss onset occurs (TGA)
V	velocity
V_{piston}	velocity of the piston during extrusion / co-extrusion (capillary-rheometer)
wt.	weight
y	gap between the rotor and the chamber wall (torque-rheometer)
Z_{ac}	acoustic impedance
η	viscosity
η_{app}	apparent viscosity
η_{int}	intrinsic viscosity
η_r	relative viscosity of the suspension
η_s	viscosity of the suspension
η_o	viscosity of the suspending medium
τ	shear stress
γ , ε	strain
$\dot{\gamma}$, $\dot{\varepsilon}$	shear rate
$\dot{\gamma}_{app}$	apparent shear rate
ϕ	volume fraction of the suspended particles
ϕ_{max}	maximum attainable concentration

Ø	diameter
ρ	density
ρ_G	initial fractional green density
ρ_s	sintered density
Ω	angular velocity
λ	wave length
d_{33}^{*}	high voltage piezoelectric coefficient

CONTENTS

1 Introduction

Lead zirconate titanate solid solution ceramics [$Pb(Zr_xTi_{1-x})O_3$; PZT] in fibre form are increasingly used to manufacture smart ceramic-polymer composites, such as rod composites (i.e. 1-3 composites) and active fibre composites (AFCs). These structures combine the high-coupling effect of PZT ceramics with the low acoustic impedance of polymers. They find their use in several types of applications, such as ultrasonic transducers (ranging from micro-speakers to medical ultrasound), health monitoring systems as well as structural control devices **[Nel02]**. Fine-scaled diameter fibres (< 250 µm) are desirable for such composites, due to the fact that they can be operated with increased resolution and actuation. To overcome the difficulty of handling such thin fibres, different methods had been developed for producing semi-finished piezoelectric ceramic-polymer composites **[Jan95b]**. Among these processing techniques, which include the injection moulding method, the tape casting technique, the "dice and fill" approach, the thermoplastic co-extrusion technology stands out due to its ability to form multiple ceramic phase composites in reduced processing steps. Additionally, devices of arbitrary geometry (complex shapes) on the micro-scale can be achieved applying this processing technique.

The ceramic micro-fabrication by co-extrusion, as a method to produce axisymmetric ceramic objects with micrometer-size features in two dimensions, was first reported by the group of John W. Halloran at the University of Michigan, USA **[Van98]** and further patented **[Hil05]**. The principle of the process consists of using a complex die, where two different extrudable mixtures (feedstock) flow together building up composite structures **[Che01]**. Alternatively, a macro-scaled preform composite is fabricated and afterwards extruded through a simple die to generate fine structures which are identical in their geometry and composition to the original preform, albeit with reduced cross sectional dimensions. Assembling the firsts extruded structures the process can be repeated until that the desired scale is achieved, restricted to the particle size of the starting materials. The main challenge of the co-extrusion method is the optimization of materials that can be co-extruded and are capable of maintaining the preform geometry in the final product without axial and cross sectional deformation. Due to this fact, no continuous thermoplastic co-extrusion of advanced ceramics is, up to date, practiced commercially.

Defect-free and homogeneous co-extruded products can only be achieved with a systematic characterisation of the processing step, where the rheology of the compositions to be co-extruded plays an important role. Differences in the flow behaviour of the materials being processed can lead to the formation of a number of instabilities depending on other parameters such as process settings and die-details. A common problem which is often reported in the polymer co-extrusion literature (**[Tak98]**, **[Doo03]**) is the encapsulation phenomena, caused mainly by the tendency of the less viscous material to migrate towards the region of highest shear since this minimises energy dissipation. Interface instabilities might additionally occur due to discrepancies in the rheology of the compositions **[Min75]**. However, in ceramics these instabilities have not been widely reported. Moreover, scientific works relating highly loaded ceramic thermoplastic co-extrusion to rheology are scarce in the literature, besides that the topic is not systematically addressed. The information available in the literature concerning co-extruded ceramic materials does not deal with the processing steps, neglecting experimental corroborations.

The primary objective of the current research was to obtain PZT solid and hollow-fibres by the thermoplastic co-extrusion process. In the course of this work, initially, in Chapter 2 the fundamentals required to understand the current study are reviewed, and Chapter 3 then details the experimental procedures used. From a processing standpoint, the work was divided into three different sections: 1) selection of the materials, 2) development, rheological characterisations and subsequent co-extrusion of the feedstocks, and 3) sintering and characterisation (microstructure and electromechanical properties) of the co-extruded piezoelectric materials. These three aforementioned steps are described in the combined results/discussions Chapters 4, 5 and 6, respectively. Chapter 7 concludes this thesis with a summary of the relevant aspects defined at each step of the work and a discussion of possible future studies. This research may function as a detailed basis for further studies to attain complex shaped ceramic materials through the co-extrusion process.

2 Fundamentals

The aim of this chapter is to present the fundamentals required for understanding the current work. In view of the objective of the present study, the chapter is divided into two sections: section 2.1 - Piezoelectric ceramic fibres and, section 2.2 - Thermoplastic co-extrusion process. Initially, the essentials of piezoelectricity as well as the general characteristics of PZT-based materials are presented. The use of PZT structures in fibre form are then discussed considering the performance of 1-3 composites. Some relevant PZT -fibres and -composites manufacturing methods are then described. Subsequently, the extrusion in general and the thermoplastic co-extrusion process in particular are reviewed. Finally, the role of rheology, a prerequisite science to develop the co-extrusion technique, is considered.

2.1 Piezoelectric ceramic fibres

2.1.1 Fundamentals of piezoelectricity

Piezoelectric materials are those in which an applied stress results in the generation of a voltage across the crystal (direct piezoelectric effect), or conversely, an applied electric field induces a change in lattice constant (converse piezoelectric effect) **[Jaf71]**. These effects are exhibited by a number of naturally occurring crystals, for instance quartz, tourmaline and sodium potassium tartrate. The crystal-symmetry plays a decisive role in the piezoelectric effect; for a crystal to exhibit piezoelectricity, its structure should have no centre of symmetry. As shown in Figure 2.1, of the 32 crystal-classes (or point groups), 11 possess a centre of symmetry, making piezoelectricity a null property for such materials. For those, an applied stress results in symmetrical ionic displacements so that there is no net change in dipole moment. The other 21 crystal classes are non-centro-symmetric, and 20 of them exhibit the piezoelectric effect. The single exception, in the cubic system

432, the combination of symmetry elements eliminates piezoelectricity. Of those 20 point groups, 10 are polar, that is, they have a vector direction in the material that is not symmetry-related to other directions. Such materials can have a spontaneous polarization, which is typically a function of temperature. Thus, these materials are pyroelectric. While symmetry considerations can describe whether a material is potentially piezoelectric, they provide no information on the magnitude of the piezoelectric response. Insight into the latter is provided via structure–property relationships that require a combination of crystallographic information with functional property data **[Mck08]**.

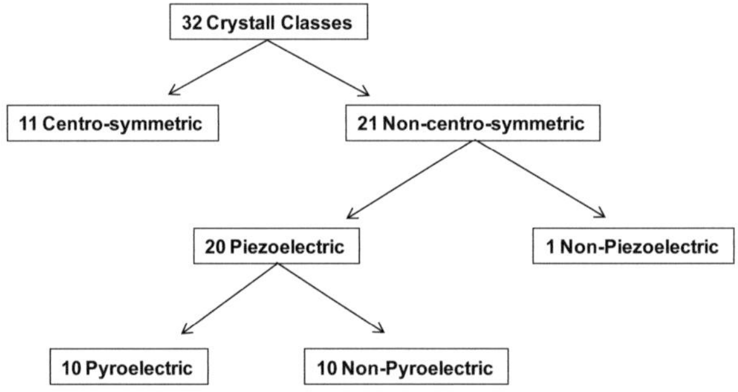

Figure 2.1: Symmetry hierarchy for piezoelectricity **[Mck08]**.

In addition to the direct and converse piezoelectric effect, a sub-group of piezoelectrics exhibits a spontaneous polarisation ($[P_s]$ or the relative displacement of ions). Since the direction of the polarisation can be reversed upon the application of an electric field (E), those materials are termed ferroelectrics on the analogy to ferromagnetic materials. All ferroelectric materials are both piezoelectric and pyroelectric. The spontaneous displacement of ions relative to each other creates permanent dipole moments. The density of such dipole moments is termed polarisation, which is a vector **[Akd08]**.

Because there are multiple possible directions for the spontaneous polarisation, in different volumes of material the polarisation nucleates in different orientations. Following, a *domain* can be defined as a volume of material in which the direction of the spontaneous polarisation is uniform (or at least nearly so). The number of allowed domain states is given by the number of possible polarisation directions. Upon application of a sufficiently high electric field, ferroelectric domains can be oriented in a process called "poling". While domains cannot be perfectly aligned with the field except when the grain or crystal is coincidentally oriented with its *c-* or *a*-axis in the field direction, their polarisation vectors can be aligned to maximize the component resolved in the field direction. *Domain walls* demark the boundaries between domains and, the phenomenon wherein the ferroelectric material changes from a spontaneously polarised state to another under electrical or mechanical loads is named *domain switching* (however, as mentioned, what is switched is the polarisation vector) **[Mou03]**. A consequence of the resistance to domain switching is that polarisation in a ferroelectric is hysteretic; it is not precisely reversible with field. This behaviour is illustrated in Figure 2.2 (a). Similar behaviour is observed in the strain (S) hysteresis, commonly referred to as butterfly loop (Figure 2.2 (b)).

The extent of the reorientation of domains along the applied field direction is governed by the point group (crystal-classes) of the crystal in question. Once all the domains have "switched", further increase in the applied field results in a proportional increase in the polarisation until a saturation value is reached (P_{sat}). Upon the removal of the applied field, the ferroelectric does not revert to its initial state although some domains do switch back. The value of the polarisation at zero field after poling is called the remnant polarisation (P_r). The subsequent application of a uniform field along the negative *z*-axis increases the volume fraction of domains with $P = -P_s$. At a critical field $-E_c$, called the coercive field, the volume fractions of the domains become equal and the net polarisation is zero. Again, an increase in the negative field eventually yields a single-domain state with the polarisation pointing in the negative *z*-direction. Upon removing the negative field, the polarisation of the ferroelectric reverts to $-P_r$. The hysteresis arises from the energy required to reverse the dipoles during the applied electric field. The area of the polarisation (P) hysteresis represents the energy dissipated in the polarisation process.

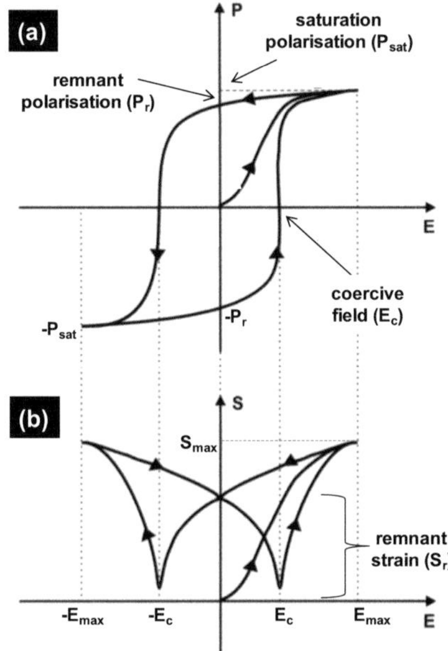

Figure 2.2: Schematic hysteresis of a ferroelectric material: (a) polarisation *P* as a function of electric field *E* and (b) strain *S* as a function of electric field *E*.

Contributions to ferroelectric switching can be divided into two components:
(1) Intrinsic contribution (or volume contribution): Response associated with the lattice properties of the material (an applied electric field causes a shift of the atoms and a change of the unit cell dimensions).
(2) Extrinsic contribution: Response associated with motion of domain walls or defects.

The aforementioned components can furthermore influence each other, as the reorientation of domain structures may stress the lattice **[Hal04]**. The represented shape of the *P-E* and *S-E* hysteresis loop (Figure 2.2) may be altered by different influences of the composition, such as the introduction of dopants, aging or fatigue **[Schö08]**.

2.1.1.1 General characteristics of PZT-based material

Commercially most important piezoelectric ceramics are based on lead zirconate titanate ($Pb(Zr_xTi_{1-x})O_3$) solid solutions. Majority of these materials were discovered in the 1950s and 1960s, and their properties and applications are described in classical textbooks (e.g. **[Jaf71]**). Its widespread use in industry, science, medicine, transportation, communications and information technologies has made PZT one of the most studied ferroelectrics. However, a limitation of PZT derived materials is the fact that they are lead-based. Lead has a high vapour pressure at elevated temperatures, making the manufacturing process complex and environmentally hazardous.

The PZT solid solution crystallizes in the perovskite structure. This structure may be described as a simple cubic unit cell with a large cation (A) on the corners, a smaller cation (B) in the body centre, and oxygen (O) in the centre of the faces, with common formula as $A^{2+}B^{4+}O_3^{2-}$ (Figure 2.3) **[Akd08]**.

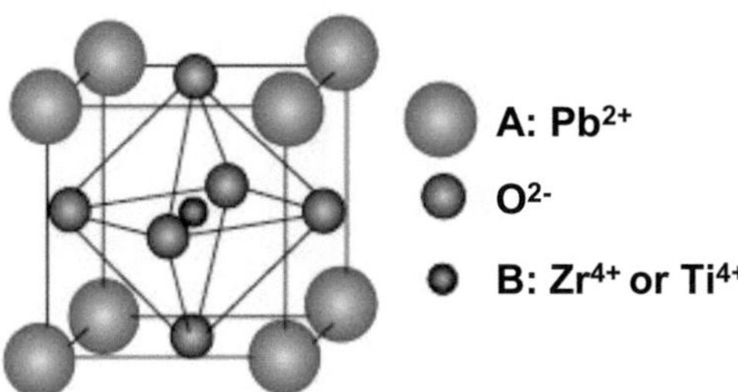

Figure 2.3: Cubic perovskite unit cell as characterized by PZT above its Curie temperature.

PZT materials are nonpolar and cubic above the Curie temperature (T_c), but become spontaneously electrically polarised and mechanically distorted below T_c. Depending on the Zr/Ti-ratio (Figure 2.4), below T_c PZT transforms into a rhombohedral or a tetragonal phase, depending on composition. The morphotropic phase boundary (MPB) defines the composition where both phases coexist. In the $Pb(Zr_xTi_{1-x})O_3$ solid solution, the MPB occurs at around $x \sim 0.535$ (Figure 2.4) where the electromechanical properties indicate a maximum.

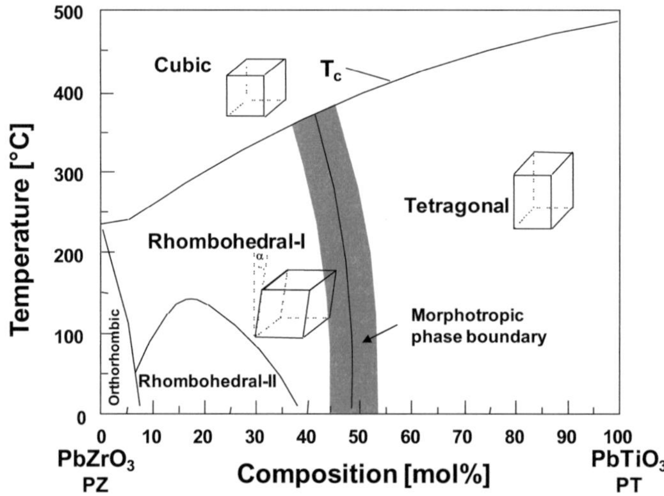

Figure 2.4: PZT phase diagram (adapted from **[Jaf71]**). The grey shaded area shows the coexistence region of the tetragonal and rhombohedral phase.

Figure 2.5 compares the butterfly loops for rhombohedral, morphotropic and tetragonal PZT materials, pointing out the higher performance of morphotropic compositions. The increased difficulty in domain switching for the tetragonal PZT (Figure 2.5 (c)) is a result of the high lattice distortion of these materials, which results in a clamping of domains **[Hof01]**.

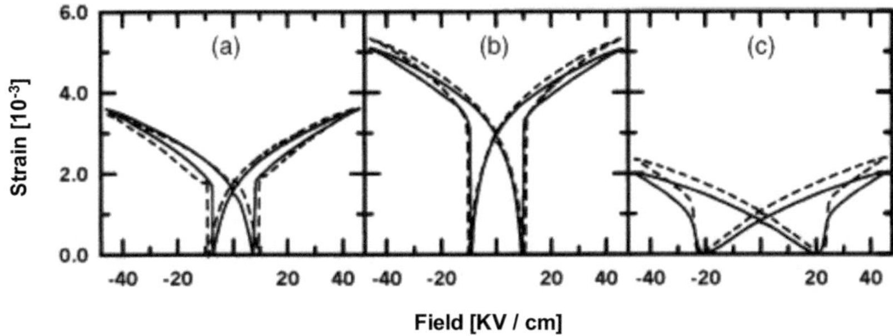

Figure 2.5: Strain hysteresis for (a) rhombohedral, (b) morphotropic, and (c) tetragonal coarse-grained (3.5 µm; solid lines) and fine-grained (1.5 µm; dashed lines) PZT **[Hof01]**.

PZT ceramics are rarely used as chemically "pure." Commercially available powders are prepared with dopants and additives incorporated into the perovskite structure in order to tailor the properties for specific applications **[Mou03]**. Substitution can take place using isovalent (e.g., Ba^{2+}, Ca^{2+}, Sr^{2+} on A-sites and Sn^{4+} on B-sites) or aliovalent ions with higher or lower valencies. The common aliovalent dopants in perovskite-type ceramics are listed in Table 2.1. Donor dopants or "soft-dopants", i.e. those of higher charge than that of the ions they replace, are compensated by A-site vacancies in the crystal structure, which enhances domain reorientation. Acceptors or "hard dopants", i.e. dopants of lower charge than that of the replaced ions, are compensated by oxygen vacancies, which reduce domain mobility due to the formation of additional dipoles.

Table 2.1: Common aliovalent substituents **[Mou03]**

Soft dopants	A-site donors	La^{3+}, Bi^{3+}, Nd^{3+}
	B-site donors	Nb^{5+}, Ta^{5+}, Sb^{5+}
Hard dopants	A-site acceptors	K^{+}, Rb^{+}
	B-site acceptors	Co^{3+}, Fe^{3+}, Sc^{3+}, Ga^{3+}, Cr^{3+}, Mn^{3+}, Mn^{2+}, Mg^{2+}

Tailoring the PZT compound with dopants or even with complex perovskite compounds, one succeeds in synthesizing multiple component systems exhibiting unusual characteristics. An increase of the electromechanical effects and piezoelectric coefficients can be reached, as well as lower sintering temperatures, low loss of PbO, lower porosity, higher density, and finer and more uniform grains. This is due to the fact that heterovalent ion combinations (double substitution of acceptor and donor ions) favour the creation of vacancies at both A- and B- sites. This promotes the diffusion process during sintering and consequently an optimum ceramic microstructure and optimum physical characteristics **[Hel08]**.

For the preparation of doped PZT-based powders, the modifying elements are added as oxides or carbonates into the composition during mixing of the components. Usually, the concentration of the modifiers is varied from 1 to 5 moles-. %. In general, good piezoelectric properties of sintered PZT can be achieved only if the ratio Zr/Ti within the grains and from grain to grain near the MPB is nearly constant.

2.1.2 PZT-based fibres: applications

PZT materials in fibre form are suitable for incorporation into smart ceramic-polymer composites. The main advantage of the use of fibres instead of monolithic PZT ceramics is the attained improvement in the design flexibility of the composites. Smart materials are materials that receive, transmit, or process a stimulus and respond by producing a useful effect that may include a signal that the materials are acting upon it **[Tul02]**. Examples of these composites are rod composites **[Jan95a]** (i.e. 1-3 composites, Figure 2.6) and active fibre composites (AFCs) **[Ben97]**. 1-3 composites consist of piezoelectric solid fibres (active phase) embedded in a passive epoxy matrix.

Figure 2.6: Sketch of a 1-3 piezoelectric fibre composite material. Fibre spacing, fibre diameters, and their variations are defined by the applied technology **[Sch08]**.

AFCs typically comprise a monolayer of uniaxially aligned piezoelectric fibres embedded in a polymer matrix between two interdigitated surface electrodes through which the driving voltage is supplied. These composites (AFCs and 1-3 composites) find their use in several types of applications, such as transducers (ranging from micro-speakers to medical ultrasound), health monitoring systems, energy harvesting devices, vibration suppression, acoustic control, as well as structural control devices (**[Gur94]**, **[Ben00]**, **[Nel02]**, **[Bru05]**).

Referring back to the AFC structures, their major drawback is the requirement of the electric field to pass through the composite matrix. Due to the placement of the electrode on the matrix surface, electric field losses require significant high voltages (on the order of kV) for actuation. Thus, an alternative approach to solid fibre AFC's is based upon *hollow piezoelectric fibres* (**[Fer96]**, **[Bre04]**). These fibres, individually electroded on both the inside and outside surfaces, are activated by an electric field applied directly across the walls of the fibre, generating longitudinal strain due to the piezoelectric d_{31} mode. Even though the longitudinal strain is considerably decreased by using d_{31} in place of the d_{33} mode used in AFCs containing solid fibres, the required voltage can be decreased by a factor of 10, since the electric field is applied only across the wall of the fibre instead of through the matrix, thereby eliminating field losses **[Bre04]**.

2.1.2.1 Performance of 1-3 composites for ultrasonic transducer applications

As already mentioned, one of the applications of piezocomposite transducers is in ultrasonic medical imaging, an indispensable diagnostic tool that span the 1-30 MHz frequency range **[Gur94]**. The popularity of this device lies in its ability to produce real-time, high resolution three-dimensional images of internal soft body tissue without the use of potentially hazardous ionizing radiation such as X-rays **[Let08]**. The process utilizes an electromechanical transducer operating in the active, or pulse-echo mode, to transmit ultrasonic pulses into the body and to receive the weak echoes (reflection of sounds) produced by the reflections from internal structures.

The overall performance of an ultrasonic system is mainly determined by the transducer characteristics **[Let08]**. The most important material parameters for transducer applications are the effective electromechanical coupling coefficient K_{eff} of the main vibration mode, and the acoustic impedance Z_{ac}. The capability of the material to convert electrical energy into acoustical energy (or vice-versa) in a short time is quantitatively described by K_{eff}. The acoustic impedance (Z_{ac}), which is expressed as sound velocity (c) x density (ρ), should be as close as possible to that of the propagation medium, i.e. biological tissues for the case of medical imaging, in order to raise the transmission efficiency. In this context, 1-3 composites offer the best compromise for medical diagnosis. The 1-3 notation specifies the connectivity pattern for a particular arrangement of fibres and the surrounding matrix material **[New78]**, i.e., in 1-3 composites, PZT fibres are continuous in one direction (z-axis), while the polymer phase is self-connected (continuous) in all three directions (Figure 2.6). This spatial scale reduces the transverse campling effect in the active element and shows higher dielectric constant in the thickness mode even with 20 to 30 vol. % of ceramics in the composite. Additionally, these structures combine the high K_{eff} of PZT ceramics with the low acoustic impedance of polymers (Z_{ac} of polymers are in the range of that of biological tissues, 1.5×10^6 Kg/s·m^2) **[Jan95b]**. Considering medical imaging applications, the attainable resolution of the transducer is limited by the wavelength of the ultrasonic pulse transmitted into the body. The relationship between the pulse wavelength (λ) and its frequency (f) is given by:

$$\lambda = c / f \tag{2.1}$$

where c, in the human body, is approximately 1500 m/s. Equation (2.1) shows that increasing the composite's frequency of oscillation, the wavelength is decreased, thereby increasing the resolution of the transducer. In other words, the resolution in which ultrasonic waves are transmitted or received is a capability of how two points can be resolved and displayed during transmitting and receiving the pulse waves, that is, the shorter the pulse width, the higher the resolution.

Increased resolution, however, must not come with a loss of homogeneity or a gain in spurious modes ("unwanted response") of behaviour (**[Jan95b]**, **[Akd05]**). Composite transducers are typically used at their thickness resonant mode frequency. Their performance is adversely affected by cross-talk ("undesired effect") in the composite structure. The cross-talk comes from the periodicity of the structure. For a given periodicity, there is a spectrum of resonant standing Lamb waves (elastic waves), which are located at wave numbers $n\pi / d$, where d is the periodicity of the ceramic phase, and n = 1, 2, 3,…In a 1-3 composite, the values of d are shown in Figure 2.7 for unit cell edge (f_{l1}) and diagonal (f_{l2}) rods **[Aul89]**. These standing wave resonances detract from the thickness resonance (f_t) if they are located too close to it. If, however, the Lamb wave resonances are at least two times higher than the thickness mode oscillations, then the thickness resonance vibrations are uniform in the plate. The frequency of the Lamb wave resonances increases with finer periodicity (as f_{l1} and f_{l2} decrease), while the thickness resonance is only a function of the plate thickness (f_t).

Thus, fine-scaled phases result in the composite behaving as a uniform medium in the region of its fundamental thickness-mode frequency.

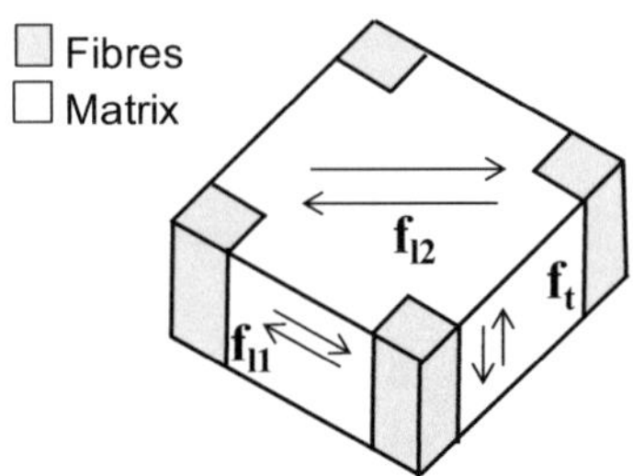

Figure 2.7: Thickness and surface stop-band modes in 1-3 composites (adapted from **[Aul89]**).

In summary, to operate the piezoelectric fibre composites at higher frequencies, for a constant volume fraction of fibres in the composite, the fibre diameter and consequently the periodicity of the lattice in the composite structure have to be reduced.

2.1.3 PZT-based fibres: fabrication methods

Processing and optimisation of piezoelectric fibres is an active research area, from which three principle manufacturing routes have emerged: extrusion **[Str99]**, sol-gel **[Mey98]** and suspension spinning **[Fre98]**. These production methods differ mainly in the precursor materials, whether a mixed oxide powder (extrusion and suspension spinning) or wet-chemical (sol-gel) route is selected. This affects stoichiometric control and the resulting properties of the PZT solid solution (**[Kor04]**, **[Den05]**). In addition, processing conditions have a great influence on the microstructure (dimensional stability, preferred orientation, defects) and on the mechanical properties of the fibres (**[Kor04]**, **[Dit10]**). Although the sol-gel process accounts for fibres diameters down to 20 µm **[Han04]**, they undergo high shrinkage after sintering, due to low powder loadings during processing. Fibres derived from the spinning process (fibre diameter between 15 and 240 µm) have the disadvantage of presenting an inhomogeneous microstructure **[Kor04]**. The thermoplastic extrusion process has proven successfully for producing fibres down to 100 µm (**[Hei05]**, **[Hei07]**). However, to obtain inferior diameters with high-solid loadings, die blockage may occur, besides that the pressure necessary to extrude would increase dramatically.

2.1.4 1-3 composites: fabrication methods

To overcome the difficulty of handling and tailoring the distance between the thin and brittle fibres, different methods had been developed for producing semi-finished 1-3 piezoelectric ceramic-polymer composites for medical imaging applications **[Jan95b]**. An overview of the most notable manufacturing methods is presented in Table 2.2.

Among these processing techniques, the thermoplastic co-extrusion technology stands out due to its ability to form multiple ceramic phase composites in reduced processing steps. Additionally, devices of arbitrary geometry (complex shapes) on the micro-scale can be achieved applying this technology **[Van98]**. However, the main challenge of the co-extrusion method is the optimization of materials that can be co-extruded and are capable of maintaining the preform geometry in the final product without axial and cross sectional deformation.

Table 2.2: Overview of fine-scaled 1-3 piezoelectric-ceramic-polymer composite processing

Method	Description/Steps of the process	Fibre diameter [µm]	Key limitations	Ref.
Dice & Fill	A series of parallel cuts are made into a block of fired PZT ceramic. After the first series of cuts are made, the block is rotated 90°, and a second series of parallel cuts are made. The rows of rods, still attached to the ceramic base, are backfilled with polymer, and the filled composite is separated from the base.	100	The width of the dicing saw blade limits the spacing of the poles.	**[Sav81]**
Relic	An alkoxide PZT stock solution is prepared and further impregnated with a woven activated carbon template material. The carbon template is then heat-treated to remove the carbon, leaving a relic in the same shape of the original template. The relic is densified by sintering and further it is backfilled with a polymer.	50	Composites can only contain a single ceramic phase, since the sol-gel solution infiltrated into the template is limited to one phase.	**[Wal92]**
Injection Moulding	PZT powder is mixed with a thermoplastic binder and injection moulded on standard equipment into preforms containing hundreds of rods. The preforms are then heat treated to remove the binder, sintered, and finally filled with a polymer.	100	Lengthy and relatively expensive process of forming the injection moulding mould.	**[Gen94]**

Tape Lamination	Tape cast sheets are stacked with a spacer support placed between each layer, and filled with a passive polymer. 1-3 composites are created by dicing the stacked sheets perpendicular to the thickness direction of the tapes.	200	The challenges in forming and handling the flat, fully dense ceramic tapes.	[Jan95a]
Fused Deposition	A thermoplastic filament feedstock is drawn from a spool and fed into a heated liquefier via a pair of counter-rotating rollers. The liquefier extrudes a continuous road of material through a nozzle and deposits it onto a fixtureless platform. The liquefier movement is computer controlled along the x and y directions. When deposition of the first layer is completed, the fixtureless platform indexes down, and the second layer is built on top of the first one. The process continuous until the manufacture of the part is completed.	100	Large-scale production is not cost-effective.	[Ban97]
Co-Extrusion	PZT powder is mixed with a thermoplastic binder and extruded to rod geometry. The rod is laminated with a carbon-filled thermoplastic sheet, and co-extruded to reduce the size of the two phases. The extrudate may be bundled together and extruded to further reduce the phase sizes.	20*	Control of the periodicity	[Van98]

* Theoretical value

2.2 Thermoplastic co-extrusion process

The general principle of the co-extrusion process can be described as the simultaneous extrusion of two or more materials through the same die. The co-extrusion is a well developed technology in the polymer field. It has been widely used to produce multilayered thin films or sheets suitable for various products, ranging from food packaging materials to reflective polarisers **[Lev89]**. The advantage of this technique in the polymer field is that it combines the desirable properties of multiple polymers into one structure with enhanced performance characteristics. Considering the co-extrusion method as a multi-pass extrusion process, at first, the subsequent subsection reviews the fundamentals of the ceramic extrusion method.

2.2.1 Ceramic extrusion: process description and mechanics

Extrusion involves forcing a deformable material through the orifice of a rigid die in order to shape a green body of semi-continuously or continuously cross-sectional area **[Ben93]**. The material to be extruded, termed extrusion paste or feedstock, is fed into the rear of the extruder and transported along the barrel either by a screw (screw or auger extruder) or by axial motion of a piston (ram or piston extruder) (Figure 2.8).

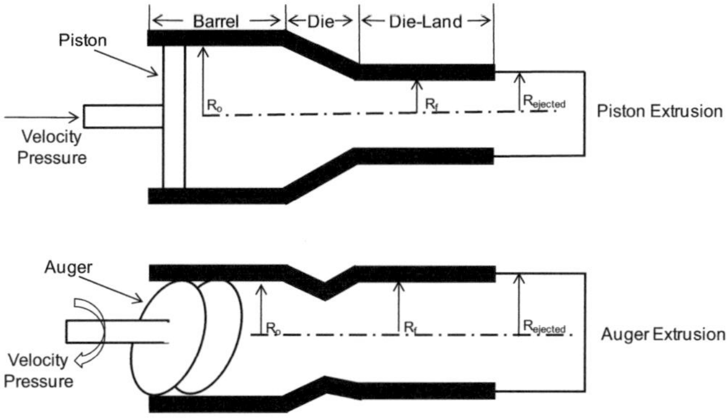

Figure 2.8: Schematic diagram of (top) piston extruder and (bottom) auger extruder indicating design parameters (adapted from **[Ree95]**).

The extrusion process includes the following stages: (a) material feeding, (b) consolidation and flow of the feed material in the barrel, (c) flow through a tapered die or orifice, (d) flow through the die-land of constant or nearly constant cross-section, and (e) ejection **[Ree95]**. The process is widely used to produce components having a uniform cross-section and a large length-to-diameter ratio such as rods, fibres, pipes, hollow tubes, and honeycombs structures. The length of the items is established by cutting the extruded material at right angles to the direction of the outflow.

This shaping method is applied to a wide spectrum of materials including monolithic metals and polymers, pharmaceuticals, foodstuffs, agricultural chemicals, as well as metal and ceramic powders. In case of monolithic metals and polymers, the material is brought to flow exclusively by the application of suitable temperatures and pressures. In the cases which involve extrusion of particulate materials, a plastically deformed vehicle (or binder phase) is applied to the material which imparts the desired flow properties. Depending on the nature and application of the extrudate, the binder phase may subsequently be removed after the shaping process (debinding or binder removal step).

During extrusion, the driving force produced by the auger or piston must exceed the resistive force of the plastic composition and the die-wall friction. Flow during extrusion may occur by one of several mechanisms **[Ree95]**:

1. Laminar flow within the extrudate and slippage at the wall (wall slip effect). This type of flow commonly occurs in the barrel and in the die-entry regions.
2. Plug flow of the extrudate and slippage at the wall (wall slip effect). This type of flow commonly occurs in the die-land.
3. Plug flow near the centre and laminar flow near the wall. This type of flow may occur when the yield point of the extrudate is very low.

These types of flow are illustrated in Figure 2.9. On flowing into the die, the material is accelerated and translates toward the central axis. The particular flow behaviour depends on the geometry of the die and the flow properties of the material. Internal defects, surface smoothness, and particle orientation are dependent on the flow behaviour during extrusion.

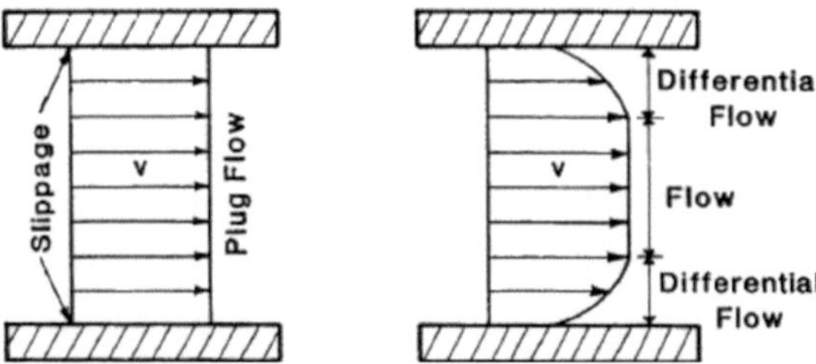

Figure 2.9: Cross-section velocity profiles for (left) plug flow with slippage at the wall and (right) differential laminar flow near the wall **[Ree95]**.

2.2.1.1 Binder phase for ceramic powder extrusion

The first requirement for the ceramic extrusion process is to attain a feedstock with the desired plasticity. Originally, for the extrusion of ceramic bricks and tiles, clay and water were used with this purpose. However, high-performance ceramics, as PZT materials, often require the elimination of clay from extrusion formulations due to the fact that the chemistry of the clay is incompatible with that of the ceramic. Hence, to provide plastic flow, the organic vehicles often used when working with high-performance ceramics are thermoplastic materials (**[Weg98]**, **[Cle07a]**, **[Bla08]**).

In addition to provide the plasticity and modify the rheological behaviour of the ceramic feedstock, the thermoplastic binders confer handling strength and perform as a backbone material to the as-shaped ceramic bodies. Their presence, however, creates significant challenges, especially for components such as filaments and thin structures. The binder must show thermal stability under mixing and shaping conditions because binder degradation can drastically change the flow properties of the feedstocks and therefore influence the geometrical properties of the final product. For practical reasons, it must exhibit high flexibility and a relatively low melt viscosity, particularly when mixed with high solid loading of ceramic powders (50 – 60 vol. %). The melt viscosity of a polymer at a given temperature is a measure of the rate at which chains can move relative to each other **[She99]**. This will be controlled by the ease of rotation around the backbone bonds, i.e., the chain flexibility, and the degree of entanglement.

The thermoplastic materials are known as viscoelastic materials, responding to external forces in a manner that is intermediate between the behaviour of an elastic solid and a viscous liquid **[Bry99]**. As a consequence, not only the viscous part of the composition will be relevant for the extrusion process. The elastic forces performing on the materials during extrusion additionally have a significant effect. For example, a common viscoelastic effect is the die swell (or extrudate swell), which happens due to a partial recovery of the elastic deformation that a viscoelastic fluid undergoes during flow through a capillary. Moreover, the thermoplastic binder must confer adequate strength to the body during the initial stage of binder removal and thermally decompose at a relatively low temperature in order not to affect the densification of the ceramic. After shaping, the binder must be removed before the sintering of the ceramic powder takes place, which is often complex and time consuming. For example, carbon residue from the organic species or uncontrolled decomposition reactions may generate defects like cracks or may entrap air and/or CO_2 during binder removal **[Lew97]**.

Although a proper binder is essential, it is not the only requirement when working with high solids loading ceramic-filled thermoplastic compounds. A surface active agent (e.g. surfactant) may be added to the system in order to ensure that the binder wets and deflocculates the powder particles and that no phase separation between powder and binder occurs under pressure. In addition, plasticizers and coagulants can be added to the system in order to modify the rheology of the feedstocks.

The surfactant performs a particularly critical function in that it forms the interface between the ceramic powder and the binder phase and thus must be compatible with both entities. To fulfil this requirement, surfactant molecules generally consist of two molecular groups with different functionalities (e.g. polar and non-polar). Thus, one end of the molecule attaches either physically or chemically to the ceramic surface and the opposite end is soluble in the thermoplastic binder **[Ree95]**. The bond with the ceramic surface must be strong enough to prevent desorption of the surfactant molecule under the shear forces encountered during extrusion. Additionally, the use of a surfactant is normally necessary to stabilize the mixture against agglomeration (functioning as a dispersant). Dispersion of ceramic powders can be accomplished in a number of different ways, and the means in which dispersion takes place varies from system to system. In the case of thermoplastic extrusion, stabilization is usually carried out via steric stabilization **[Mcn99]**. In this approach, adsorbed organic molecules (often polymeric in nature) are utilized to induce steric repulsion. To be effective, the adsorbed layers must be of sufficient thickness and density to overcome the van der Waals (vdW) attraction between particles and to prevent bridging flocculation.

Such species should be strongly anchored to avoid desorption during particle collisions **[Lew00]**.

2.2.2 Thermoplastic co-extrusion in the ceramic field

The ceramic micro-fabrication by co-extrusion, as a method to produce axisymmetric ceramic objects with micrometer-size features in two dimensions, was first reported by Van Hoy et al. at the University of Michigan, USA **[Van98]**. A PZT-thermoplastic compound was prepared at 50 vol. % ceramic loading using a high-shear mixer. The obtained feedstocks were formed into sheets 0.3 mm thick by pressing between steel plates. A fugitive compound (a material that functions as space filler between the green filaments, provided that it can be filled with a thermoplastic material and completely removed after the micro-fabrication) was similarly pressed into 0.22 mm thick sheets. These were cut into squares and stacked alternately. The stack was wrapped with PZT compound to form the first-stage feedrod. This was extruded through a square orifice at a reduction ratio of $R = 5.5$. After binder and fugitive material removal, and sintering at 1250 $^{\circ}$C, a free-standing array of dense PZT ceramics sheets was obtained. The array had regions of good order, but also regions where the lines were distorted.

An alternative co-extrusion method to the aforementioned one is available where the preform to be reduced is formed in a complex die in which continuous paste streams are brought together into the desired structure **[Lia01]**; this is sometimes called the multi-billet process **[Che01]**. The ceramic co-extrusion technology has been proven to be effective to fabricate fibrous monolithic materials (**[Kov97]**, **[Kim04]**), micro-configured multi-component ceramics **[Cru98]**, honeycomb structures **[Buc01]**, alumina fibres **[Gre01]**, biocompatible macro-channelled hydroxyapatite **[Koh02]**, multilayered actuators **[Yoo05]**, piezoelectric PZT-air composites (2-2 connectivity) **[Cru07]**, as well as dual-layer hollow fibres **[Dro09]**.

The ceramic-thermoplastic co-extrusion process developed within this work, includes the following steps: a) feedstocks preparation and rheological characterisations, b) preform composite fabrication followed by co-extrusion, c) removal of the organic vehicles (debinding) and, finally, d) sintering of the body to near full density. In order to comprehend the characterisation methods used as well as the results/discussions of the later chapters, the subsequent subsections review the essential aspects during processing by co-extrusion.

2.2.2.1 Feedstocks preparation

The feedstocks for plastic forming are commonly prepared by directly mixing the raw materials and additives in a high-shear mixer. This step, named compounding or mixing, is a crucial stage for the co-extrusion technology, since mixing deficiencies cannot be corrected by subsequent processing adjustments. A number of variables may affect the final quality of the mixture: rotor geometry, mixing time, rotor speed, chamber loading, mixing temperature as well as the sequence of material addition.

The purpose of mixing is to attain a homogeneous composition, free of large agglomerates, to prevent flow related failures. The process involves the breaking down of the ceramic agglomerates, their separation and segregation, until a final random distribution is achieved of each component in the system. Measurement of the torque during mixing as a function of mixing time is useful for determining the effects of procedures and additives and for control purposes **[She99]**. A typical curve of the torque as a function of time is shown in Figure 2.10. The curve illustrates the thermo-mechanical history experienced by the thermoplastic binder and the ceramic under mixing. When the thermoplastic binder is introduced in the heated mixing chamber, it offers a certain resistance to the free rotation of the blades and therefore an increase of the torque is verified (first maximum, Figure 2.10). This resistance is overcome when the heat transfer is sufficient to completely melt the thermoplastic material. Thus, the torque required to rotate the blades at the fixed speed decreases and reaches for a short time a steady state. At this stage, the ceramic powder is gradually loaded. An increase of the torque occurs up to that a second maximum is achieved, an indicative of the energy necessary to fill voids between particles and cover their surface. This value points out the force required to destroy the agglomerates present in the ceramic material. When the ceramic surface is completely covered by the thermoplastic binder, the torque decreases and reaches again a steady state regime. The time required to reach the steady state depends on the material and on the processing conditions (temperature and rotor speed). A stable torque is a qualitative indication that the filler is uniformly distributed in the polymer binder. Consequently, at this point, the composition is assumed to be homogeneous. Mixing longer, an increase or decrease of the torque might occur, depending whether cross-linking or degradation phenomena of the binder take place.

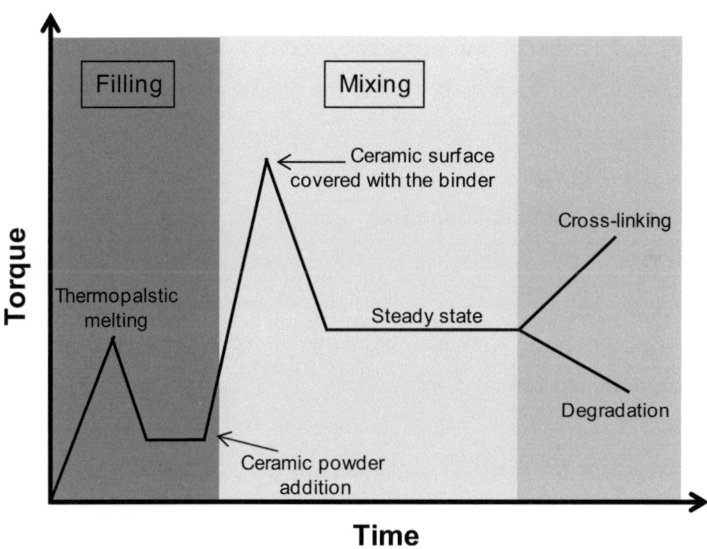

Figure 2.10: Schematic representation of the measured torque as a function of time during feedstock compounding.

2.2.2.2 The role of rheology on the co-extrusion process

Rheology is a fundamental interdisciplinary science concerned with the flow and deformation of matter. In order to understand and control any process involving the transfer of fluids it is necessary to know how the fluid behaves under different conditions (e.g. temperature, pressure, rates of strain). Particularly in the co-extrusion process, knowledge on the rheology of the compositions to be co-extruded plays an important role to achieve defect-free products. Differences in the flow behaviour of the materials being processed can lead to the formation of a number of instabilities depending on other parameters such as process settings and die-details.

This section is divided into three subsections. Initially, it introduces the basic rheological concepts necessary for the development of this work. Subsequently, some rheological problems in the polymer co-extrusion field are reviewed. Finally, since the rheological behaviour of the feedstocks is a function of solid loading, some viscosity models described in the literature which consider the influence of volume fraction of solids on relative viscosity are presented.

2.2.2.2.1 Basic rheological concepts

Rheology is concerned with the relationship between stress (defined as force per unit area), strain (defined as change in dimension per unit dimension) and time. There are three simple deformations:

(a) In simple shear the stress is applied tangentially (Figure 2.11):

$$\text{stress [N/m}^2] \quad \tau_s = \frac{F}{A} \qquad (2.2)$$

28

$$\text{strain [unity]} \quad \gamma = \frac{x}{h} \tag{2.3}$$

$$\text{shear rate [1/s]} \quad \dot{\gamma} = \frac{1}{h}\frac{dx}{dt} = \frac{v}{h} \tag{2.4}$$

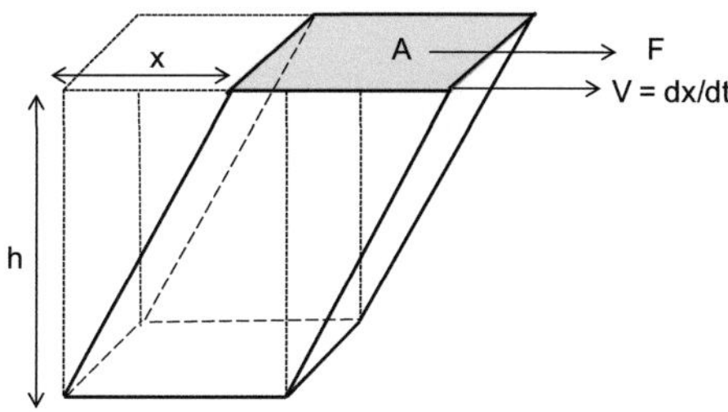

Figure 2.11: Simple shear: area A and distance h remain constant during deformation.

(b) In simple extension, the stress is applied normal to the surface of the material (Figure 2.12):

$$\text{stress [N/m}^2\text{]} \quad \tau_E = \frac{F}{A} \tag{2.5}$$

$$\text{strain [unity]} \quad \varepsilon = \int_{l_0}^{l_1} \frac{dl}{l} = \ln\frac{l_1}{l_0} \tag{2.6}$$

$$\text{shear rate [1/s]} \quad \dot{\varepsilon} = \frac{v}{l} \tag{2.7}$$

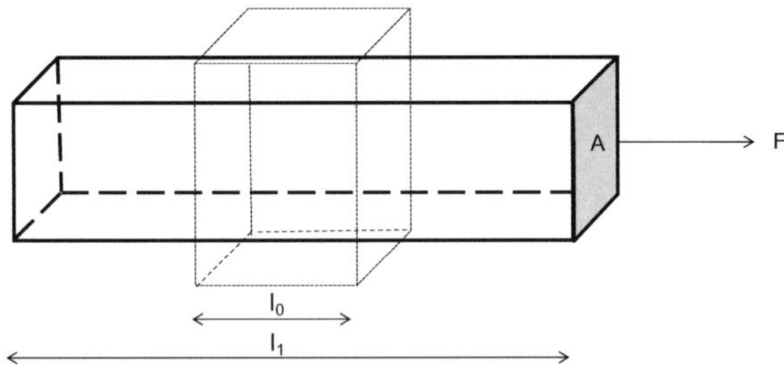

Figure 2.12: Simple extension: cross-sectional area A and sample length l both vary during deformation.

(c) In bulk deformation, the stress is applied normal to all faces. The stress is the applied pressure, P, and the strain is the change in volume per unit volume $\delta V/V$.

The practical deformations in polymer processing are themselves usually complex flows compounded of shear, extension and bulk deformations **[Cog81]**.

There are three types of response to an applied stress: viscous flow, elastic deformation and rupture. In viscous flow a material continues to deform as long as the stress is applied and the energy put in to maintain the flow is dissipated as heat. Viscosity (η) represents the resistance of a material to flow, due to the internal friction between the molecules of the material. It is defined as the ratio of shear stress to shear rate and, in the SI system, has the unit Pa·s (equation (2.8)).

$$\eta = \frac{\tau}{\dot\gamma} \qquad (2.8)$$

If the only response of a material is to flow with a viscosity which is independent of stress level, the material is said to behave as a Newtonian liquid. If the material deforms instantly under stress and the deformation is spontaneously reversed when the stress is removed, the material possesses an elastic response. If all the deformation is reversible and the deformation is proportional to the applied stress the material is said to have a Hookean response. Thermoplastic binders, as previously mentioned, demonstrate a viscous-elastic response to stress. One approach to measure the rheology of these materials is to assume that the material will respond as a Maxwell model, in which a Newtonian viscous damper is placed in series with a Hookean elastic spring **[Hea97]**. Thus, an *apparent viscosity* and an *apparent modulus* which depend on factors such as temperature, pressure, stress, geometry of deformation, as well as time, are measured **[Mac94]**.

Rheometry is the measuring technology used to determine rheological data. Instruments which measure the rheological properties of the materials are described as rheometers. Rheometers used for filled polymer systems can be divided into two broad categories - (a) rotational/oscillation and (b) capillary type. **[She99]**. In rotational instruments, a shear stress in a material is generated between a moving and a fixed solid surface (gap), whereas in capillary rheometers the shear stress is generated by squeezing a material through a capillary die.

Flow behaviour is represented graphically using flow curves and viscosity curves. The flow curve displays the mutual dependence of shear stress and shear rate, whereas the viscosity curve is derived from the flow curve. Figure 2.13 represents the viscosity curve of a pseudoplastic (or shear-thinning) material. This is a remarkable behaviour of thermoplastic materials. At very low shear rates (stage I, Figure 2.13), pseudoplastic materials behave similarly to Newtonian liquids, having a defined viscosity independent of shear rate - often called the "zero shear viscosity". This is due to the fact that the Brownian motion of molecules keeps all molecules or particles at random in spite of the initial effects of shear orientation. A new phenomenon takes place when the shear rate increases to such an extent that the shear induced molecular or particle orientation by far exceeds the randomizing effect of the Brownian motion, decreasing the viscosity drastically (stage II, Figure 2.13). Reaching extremely high shear rates, the viscosity will approach asymptotically a finite constant level, since an optimum orientation has been achieved (stage III, Figure 2.13).

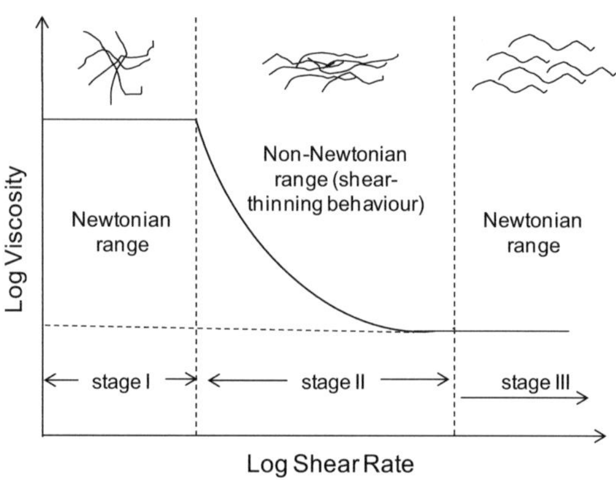

Figure 2.13: Viscosity curve of pseudoplastic materials (adapted from **[Sch04]**). The illustrations on the top represent the entanglement/disentanglement of the macromolecules as a function of the applied shear rate.

2.2.2.2.2 Rheological problems in co-extrusion

The main limiting factor in the co-extrusion technology is the flow instability that may occur during processing. A common problem which is often reported in polymer co-extrusion literature (**[Min75]**, **[Tak98]**, **[Doo02]**) is the encapsulation phenomena (or viscous encapsulation), caused mainly by the tendency of the less viscous material to migrate towards the region of highest shear, since this minimises energy dissipation. This phenomenon is illustrated in Figure 2.14. The figure shows how the less viscous layer (A) tends to move to the highest area of shear (near the tube walls) and so encapsulate the layer with the higher viscosity (B). Complete encapsulation might occur when working with extremely long dies.

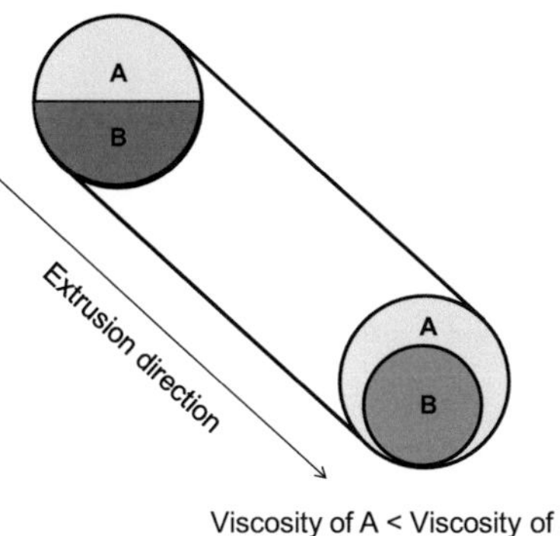

Figure 2.14: Viscous encapsulation in a tube (adapted from [Doo02]).

Differences in wall adhesion and viscoelastic characteristics of polymers are also contributing factors for the encapsulation phenomena. Weak secondary flows caused by viscoelastic effects (from the second normal stress difference) have been demonstrated to produce layer non-uniformities even in co-extrusion of different coloured streams of the same polymer **[Doo03]**. Reduction of this defect can be achieved by choosing materials with the smallest possible differences in viscosity, or by changing the stream temperatures to bring the polymer viscosities closer to one another.

Interfacial instabilities might additionally occur due to discrepancies in the rheology of the compositions (**[Zat05]**, **[Lam09]**). They represent an internal type of instability (inside the product), that is, the outside surfaces may be smooth in this case (Figure 2.15). The key to smooth parallel flow of materials is to minimize the interfacial shear stress and match the viscoelastic properties of the adjacent layers.

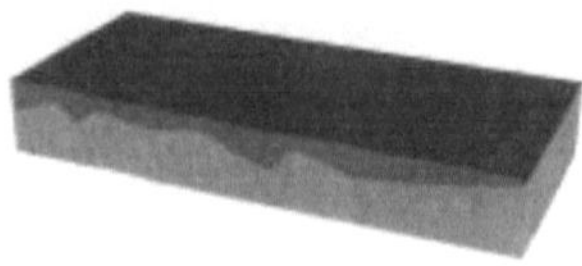

Figure 2.15: Scheme of interfacial instabilities [Zat05].

In ceramics these instabilities have not been widely reported. Beeaff and Hilmas **[Bee02]** were, so far, the only that documented rheological problems introduced during the thermoplastic co-extrusion process. Using $BaTiO_3$ and nickel based feedstocks, they verified layer displacement on multilayered architectures co-extruded when a control of the relative viscosities of the constituents was not considered.

2.2.2.2.3 Viscosity models for highly concentrated suspensions

Based on Einstein's proposed relationship between the relative viscosity of the suspension (η_r) and the volume fraction of the suspended particles (ϕ) **[She99]**, valid only for diluted solutions (equation (2.9)), a variety of empirical models have been developed to describe the effect of the solids content on the viscosity of highly concentrated suspensions.

$$\eta_r = 1 + 2.5\phi \tag{2.9}$$

where η_r is the ratio of the viscosity of the suspension (η_s) to the viscosity of the suspending medium (η_o). Such models usually consider the solids content (ϕ) the maximum attainable concentration (ϕ_{max}) and the intrinsic viscosity (η) to calculate a relative property value. Examples of these models include the Chong (equation (2.10)) **[Cho71]**, the Frankel-Acrivos (equation (2.11)) **[Fra67]**, the Quemada (equation (2.12)) **[Que77]** and the Krieger-Dougherty models (equation (2.13)) **[Kri59]**:

$$\eta_r = \left[1 + 0.75 \left(\frac{\phi / \phi_m}{1 - \phi / \phi_m} \right) \right] \qquad (2.10)$$

$$\eta_r = \frac{1}{\left(1 - \dfrac{\phi}{\phi_m} \right)^2} \qquad (2.12)$$

$$\eta_r = C' \frac{(\phi / \phi_m)^{1/3}}{1 - (\phi / \phi_m)^{1/3}} \qquad (2.11)$$

$$\eta_r = \left(1 - \frac{\phi}{\phi_m} \right)^{-[\eta_{int}]\phi_m} \qquad (2.13)$$

The following rules are considered in these models. As the volume fraction of solids, ϕ, tends to zero, the relative viscosity, η_r, tends to unity, representing the limit for the suspending medium. The volume fraction of solids tends towards the maximum attainable concentration as the relative viscosity tends to infinite. Considering particles as monodispersed spheres, the value of C' in equation (2.11) is 1.125 and that of the intrinsic viscosity η_{int} in equation (2.13) it is 2.5. The intrinsic viscosity represents the effective shape factor of the suspended particles for their movement in the shear field imposed **[Oka00]**. The maximum attainable concentration, ϕ_{max}, also known as the maximum volume loading or maximum packing fraction, is the highest filler loading possible while still retaining a continuous matrix phase. This parameter depends on filler- particle geometry and packing geometry **[Big83]**.

3 Experimental

The current chapter describes the raw materials, equipments, and experimental methods used in this work. An overview of the experimental setup is given in Figure 3.1; the details of each step will follow.

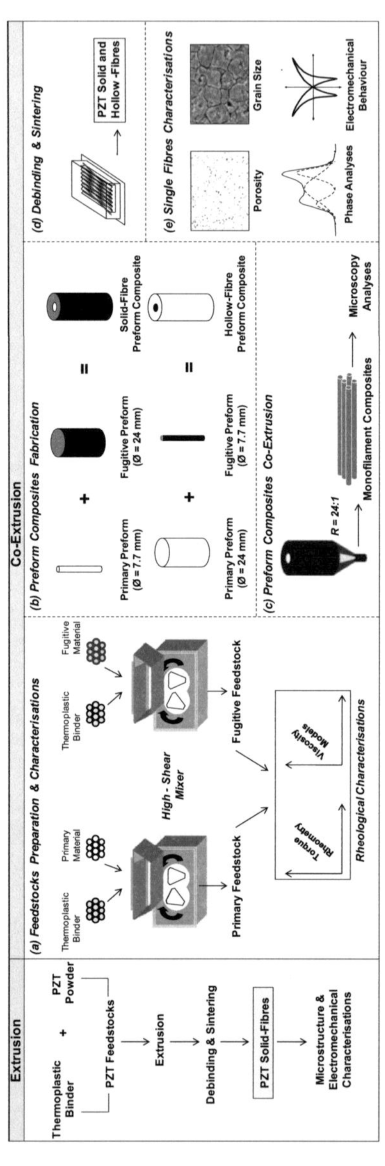

Figure 3.1: Experimental setup overview.

3.1 Raw materials: description

For the production of the solid and hollow lead zirconate titanate fibres by co-extrusion, two different feedstocks were used: one containing the primary material, which means the material to be micro-fabricated, and the other containing a fugitive material, whose function is to fill the space between the green filaments. The first step of the present work was to select the materials to compound these feedstocks. The description and function of each raw material tested are listed in Table 3.1.

Table 3.1: Raw materials used for feedstocks preparation

Raw material	Function	Product	Supplier
Lead zirconate titanate **(PZT-P505)**	Primary material	Sonox® P505	CeramTec AG, Germany
Lead zirconate titanate **(PZT-EC65)**		EC65	EDO Corporation, USA
Carbon black **(CB)**	Fugitive material	BP® 120	Cabot Corporation, USA
Microcrystalline cellulose **(MCC)**		MCC Cotton Linters	Sigma-Aldrich GmbH Switzerland
Stearic acid **(St. acid)**	Surfactant	93661 Fluka	Sigma-Aldrich GmbH Switzerland
Polystyrene **(PS)**	Thermoplastic binder	Styron™ 648	Dow Plastics, Belgium
Poly(isobutyl methacrylate) **(PiBMA)**		Acryloid B67	Lascaux, Switzerland
Low density polyethylene **(LDPE)**		PEBD 1700 MN 18C	Lacqtene Elf Atochem S.A., Switzerland
Poly(ethylene-co-ethyl acrylate) **(EEA)**		Elvaloy® 2615 AC	DuPont™, Switzerland

3.2 Raw materials: data and characterisation

The density values for all raw materials listed in Table 3.1 were obtained using a helium pycnometer (AccuPyc 1330, Micromeritics Instrument Corporation). The as-received MCC was dried for 24 h in an oven at 110 °C prior to any measurement. The particle size distribution (PSD) of the PZT and the carbon black (CB) powders was measured using a laser diffraction particle size analyzer (LS230, Beckman-Coulter Inc). The specific surface area (SSA) of these powders was determined from a five point N_2 adsorption isotherm obtained from BET (Brunauer-Emmett-Teller) measurements (SA3100, Beckman-Coulter Inc). Prior to BET analysis, the powder samples were degassed (SA-PREP, Beckman-Coulter Inc) at 60 °C for 24 h under N_2 atmosphere to remove adsorbed water from the particle surfaces. The results of these aforementioned characterisations for the PZT (PZT-P505 and PZT-EC65), CB, and MCC materials, as well as additional information, are presented in Table 3.2. For the PZT-P505, two different batches were used within this work.

Table 3.2: Characteristics of the PZT, CB and MCC materials

Raw material	Density [g/cm^3]	SSA [m^2/g]	PSD [µm]	Additional information
PZT-P505 (batch 1)	8.10	1.97 ± 0.01	d_{10} = 1.35 d_{50} = 2.66 d_{90} = 4.28	T_c = 335 °C Soft PZT
PZT-P505 (batch 2)	7.87	1.96 ± 0.05	d_{10} = 1.34 d_{50} = 2.53 d_{90} = 3.97	
PZT-EC65	7.95	1.00 ± 0.07	d_{10} = 2.24 d_{50} = 3.24 d_{90} = 5.14	T_c = 350 °C Soft PZT
CB	1.90	30.33 ± 0.03	d_{10} = 0.30 d_{50} = 1.59 d_{90} = 2.19	Produced by partial combustion of petroleum oil.
MCC	1.58	100*	-	Molecular formula: $(C_6H_{10}O_5)_n$

*Value from the literature **[Bat75]**

Figure 3.2 and Table 3.3 give further information on all of the thermoplastic binders investigated in this study. Table 3.3 additionally displays the characteristics of the stearic acid (surfactant). The EEA copolymer is composed by 85 wt. % ethylene and 15 wt. % ethyl acrylate. The composition of the blend was 75 vol. % EEA and 25 vol. % PiBMA. This selected ratio was based on the fact that greater amounts of PiBMA lead to stiffer blends, which are more suitable for large dimensions co-extruded products. On the other hand, for thin co-extruded filaments, highly flexible compounds are desired, such that the filament can resist fracturing by stretching **[Koh04]**. The melt flow index (MFI) data (Table 3.3) were supplied by the manufacturers.

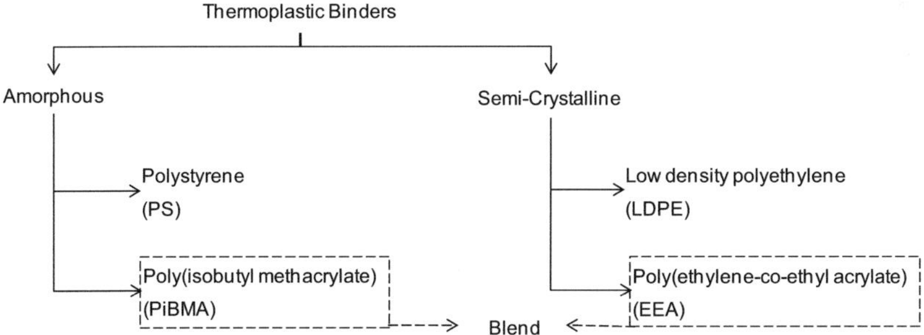

Figure 3.2: Thermoplastic binders investigated in this study.

Table 3.3: Selected information of the thermoplastic binders investigated

Organic binder	Repeat unit	$M_w{}^a$ [g/mol]	Density [g/cm³]	MFI[b] [g/10 min]
PS	—(CH₂ – CH)— (phenyl)	104.15	1.05	1.3
PiBMA	—(CH₂ – C)— with CH₃, C=O, O, CH₂, H₃C–CH, CH₃	142.20	1.04	c
LDPE	—(CH₂ – CH₂)—	28.05	0.92	70
EEA	—(CH₂ – CH₂)ₓ—(CH₂ – CH)ᵧ— with C=O, O, CH₂, CH₃	128.17	0.93	6
St. acid	H₃C–(CH₂)₁₆–COOH	284.48	0.85	-

[a] Molecular weight of repeat unit; [b] Melt flow index: 2.16 kg at 190 °C (ASTM D1238 - ISO 1133); [c] Value not supplied.

The glass transition temperature (T_g) and the crystalline melting temperature (T_m) of the thermoplastic binders were determined by differential scanning calorimetry (DSC; DSC 7 equipped with a TAC 7/DX Thermal Analysis Controller, Perkin-Elmer). The analyses were carried out from 20 °C to 160 °C, with a heating rate of 20 °C/min, under a nitrogen atmosphere. The measurement procedure included heating the sample up to 160 °C, subsequently cooling it down to 20 °C and finally re-heating it up to 160 °C. The second heating curve is the one presented in this work because this is the most appropriate for interpretation, as the thermal history of the polymer is then known. The heat flow curves were normalized for the mass of the sample.

Thermo-gravimetric analyses (TGA; TGA/SDTA 851^e STAR^e System thermobalance, Mettler Toledo) were performed by heating the organic samples (5 – 10 mg) in an alumina crucible at 1 or 3 °C/min up to 600 °C. The TG scan was conducted either in air and or in an oxygen atmosphere at a flow rate of 50 ml/min. TGA was carried out for the pure fugitive materials, surfactant, thermoplastic binders (Table 3.1) and for the developed feedstocks (Table 3.4).

3.2.1 Coating of the PZT powder

Based on the work of McNulty et al. **[Mcn99]**, the as-received PZT powders (PZT-P505 and PZT-EC65) were initially coated with stearic acid. The desired surfactant amount ($wt._{St.acid}$ [g]) was calculated based on the area occupied by one molecule given by Moore et al. **[Moo72]** and the specific surface area of the ceramic powder (SSA_{PZT} [m^2/g]) (equation (3.1)).

$$wt._{St.acid} = \frac{SSA_{PZT}.wt._{PZT}.M_{W_{St.acid}}}{Al_{St.acid}.10^{-18}.N_A} \tag{3.1}$$

where $wt._{PZT}$ corresponds to the weight of as-received PZT powder to be coated, $M_{W_{St.acid}}$ is the molecular weight of the stearic acid, $Al_{St.acid}$ is the allocation of the stearic acid (0.2 nm^2/molecule) and N_A is the Avogadro number (6.02217 x 10^{23} mol^{-1}). By applying equation (3.1), one layer of stearic acid is calculated.

Following Heiber et al. **[Hei05]**, the appropriate amount of surfactant was firstly dissolved in toluene (258 ml). Afterwards, 300 g of PZT powder was added and subsequently milled in a jar mill containing 500 g of 3 mm diameter YSZ milling balls, during 20 h. After milling, the slurry was dried during 90 min using a rotary evaporator (Rotavapor R-134, Büchi) at 30 rpm, 80 mbar and 55 °C to extract the solvent. With the goal to ensure dispersion, the particle size distribution of the uncoated (PZT uncoated) and the coated (PZT coated) powders was measured using the laser diffraction particle size analyzer mentioned before. Prior to the measurements, coated and uncoated PZT powders were dispersed in a toluene media and homogenized through ultrasonic treatment (HD 2070, Bandelin Sonopuls).

3.3 Feedstocks preparation

The feedstocks were prepared using a high-shear mixer (HAAKE PolyLab Mixer Rheomix 600, Thermo Fisher Scientific). The total volume of the compounds was formulated to fill 70% of the volume of the mixer chamber (volume of the chamber with the roller rotors = 69 cm^3). In order to improve homogeneity and reduce particle agglomeration, the mixing was carried out using a two steps sequence: first at 10 rpm for 30 minutes at a defined temperature (factor determined depending on the characteristics of the thermoplastic binder used) and subsequently at 10 rpm until the torque reached steady-state conditions. The unfilled thermoplastic binders were mixed using only the aforementioned first step mixture. The different feedstocks prepared within this work and their formulations are listed in Table 3.4.

It is worth noting that the MCC feedstocks contained stearic acid in their composition. The cellulose particles posses a strong tendency to agglomerate as they are hydrophilic and, in the case of the present study, mixed with a hydrophobic thermoplastic material. The use of this fatty acid containing both hydrophobic and hydrophilic groups (methyl and carboxyl groups, respectively) overcame this phenomenon.

Table 3.4: Feedstocks characterized within this work and their formulations

N°	Feedstocks	Formulations [vol. %]
1	Unfilled PS	100 PS
2	Unfilled LDPE	100 LDPE
3	Unfilled EEA	100 EEA
4	Unfilled Blend	75 EEA : 25 PiBMA
5	58 vol. % PZT + PS	58 PZT (P505 batch 1) : 7.7 St. acid : 34.3 PS
6	58 vol. % PZT + EEA	58 PZT (P505 batch 1) : 7.7 St. acid : 34.3 EEA
7	58 vol. % PZT + Blend	58 PZT (P505 batch 1) : 7.7 St. acid : 25.7 EEA : 8.6 PiBMA
8	52 vol. % PZT uncoated + LDPE	52 PZT (P505 batch 1) : 48 LDPE
9	52 vol. % PZT + LDPE	52 PZT (P505 batch 1) : 6.9 St. acid : 41.1 LDPE
10	56 vol. % PZT + LDPE	56 PZT (P505 batch 1) : 7.4 St. acid : 36.6 LDPE
11	58 vol. % PZT (1) + LDPE	58 PZT (P505 batch 1) : 7.7 St. acid : 34.3 LDPE
12	58 vol. % PZT (2) + LDPE	58 PZT (P505 batch 2) : 7.7 St. acid : 34.3 LDPE
13	58 vol. % PZT (3) + LDPE	58 PZT (EC65) : 7.7 St. acid : 34.3 LDPE
14	25 vol. % CB + LDPE	25 CB: 75 LDPE
15	35 vol. % CB + LDPE	35 CB: 65 LDPE
16	45 vol. % CB + LDPE	45 CB: 55 LDPE
17	20 vol. % MCC + LDPE	20 MCC : 2 St. acid : 78 LDPE
18	30 vol. % MCC + LDPE	30 MCC : 2 St. acid : 68 LDPE
19	40 vol. % MCC + LDPE	40 MCC : 2 St. acid : 58 LDPE
20	41 vol. % MCC + LDPE	41 MCC : 2 St. acid : 57 LDPE
21	50 vol. % MCC + LDPE	50 MCC : 2 St. acid : 48 LDPE

3.4 Fibre extrusion

PZT green solid-fibres were produced by extruding vertically the feedstocks containing the different thermoplastic binders. A capillary rheometer (RH7

Flowmaster, Rosand Precision Limited) was used for this purpose (Figure 3.3). The velocity of the piston was constant during the extrusion step (V_{piston} = 0.176 mm/min). The orifice of the die used was 300 µm in diameter.

The barrel of the capillary rheometer contains three heater bands in order to ensure temperature uniformity during processing. Before performing actual extrusion, the feedstocks were heated in the barrel for 20 min in order to achieve a homogeneous temperature distribution. The die swell S_w of the extruded fibres was calculated as the ratio of the extruded diameter to the orifice diameter of the die. The extruded diameter was measured using an optical microscope (Wild M3Z, Leica) equipped with a micrometer (Mikrometer Mitutoyo 25 mm).

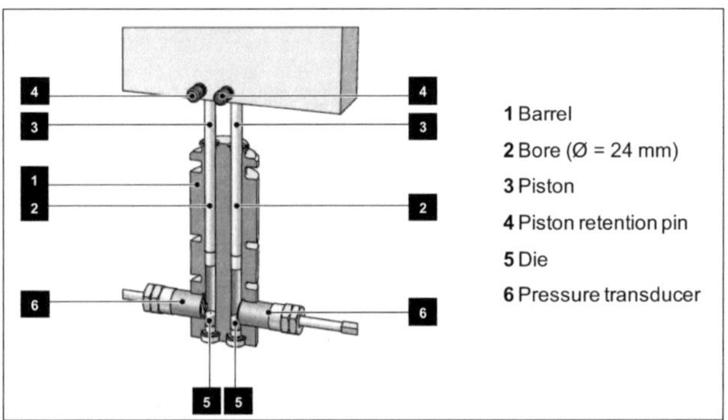

Figure 3.3: Detailed barrel area of the capillary rheometer (adapted from **[Sch04]**).

3.5 Fibre co-extrusion

The steps used for processing the different geometry-shaped fibres (solid and hollow) by co-extrusion are described in the following subsections.

3.5.1 Rheological characterisations of the feedstocks

The rheological behaviour of the feedstocks were characterised by means of torque-rheometry and in view of the viscosity equations derived by Chong, Frankel-Acrivos, Quemada and Krieger-Dougherty (Chapter 2).

3.5.1.1 Torque-rheometry

A torque rheometer (HAAKE PolyLab Rheometer, Thermo Fisher Scientific) was used to record the rheological data of the feedstocks listed in Table 3.4. A pair of parallel aligned roller rotors counter-rotating at a speed ratio of 3:2 was used (only the higher speed is indicated on the dial). A representation of the torque rheometer and its geometric dimensions are shown in Figure 3.4. After compounding (section 3.3), equilibrium torque values were recorded at rotor speeds ranging from 10 to 100 rpm with steps of 10 rpm, at different temperatures. A stable torque is a qualitative indication that the filler is uniformly distributed in the polymer binder. It is worthy to mention that adding particles to the system increased the time to achieve torque equilibrium.

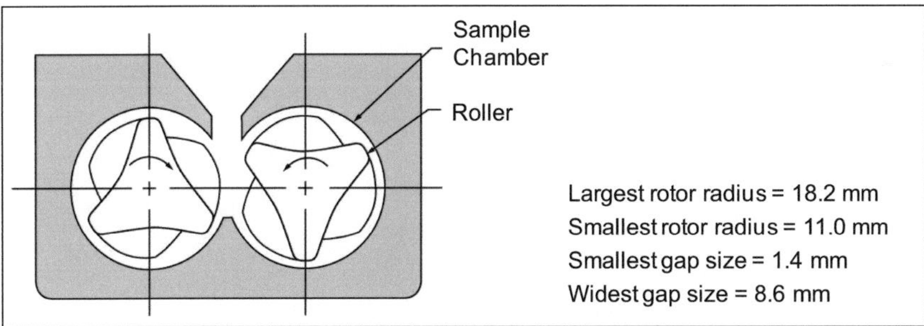

Figure 3.4: Schematic front view of the torque-rheometer and its geometrical dimensions.

3.5.2 Solid-fibre production

3.5.2.1 Preform composite assembly

As mentioned, the preform composite for micro-fabricating PZT solid *(and hollow)* fibres by the co-extrusion process is formed by two different feedstocks. The primary feedstock contained the ferroelectric ceramic (PZT-based material) which was to be micro-fabricated. The secondary (fugitive feedstock) contained the fugitive material (CB or MCC), which functioned as space filler between the green filaments.

For the solid-fibre production, in order to obtain the primary preform, a primary feedstock was extruded vertically through a 7.77 mm diameter die using the capillary rheometer defined before (Figure 3.3). The velocity of the piston was 5.0 mm/min at a defined temperature (based on the characteristics of the thermoplastic binder used). To obtain the fugitive preform, a fugitive feedstock was uniaxially pressed (OPUS, Römheld) into a cylinder shaped feedrod (diameter = 24 mm, length = 45 mm). A hole with the same diameter of the primary preform was drilled in the centre of the fugitive cylinder in order to structure the preform composites (Figure 3.5).

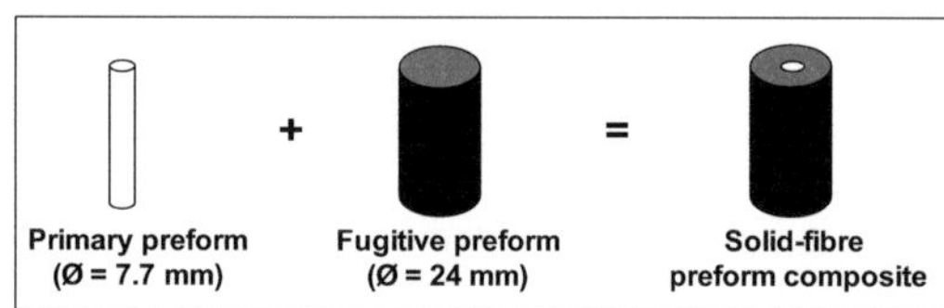

Figure 3.5: Preform composite assembly for solid-fibre production.

3.5.2.2 Co-extrusion parameters

The co-extrusion experiments for solid-fibre production were carried out using the capillary rheometer described before. The temperature was defined based on the characteristics of the thermoplastic binder material selected. A die with a half cone angle of 60°, 1 mm diameter and 16 mm length was used, so that the extrudate was 24 times smaller than the initial preform (R = 24:1). Varying the speed of the piston (V_{piston}), 5 different experiments were performed: Co-Ex 1 (V_{piston} = 0.07 mm/min), Co-Ex 2 (V_{piston} = 0.26 mm/min), Co-Ex 3 (V_{piston} = 0.78 mm/min), Co-Ex 4 (V_{piston} = 1.30 mm/min) and Co-Ex 5 (V_{piston} = 3.91 mm/min).

3.5.3 Hollow-fibre production

3.5.3.1 Preform composite assembly

The preform geometry for the hollow-fibre production was the opposite than the preform geometry for the solid-fibre production (section 3.5.2.1), i.e., the primary feedstock was uniaxially pressed into a cylinder shaped feedrod (24 mm diameter), and the fugitive feedstock was extruded through a 7.77 mm diameter die. The preform composites were then structured drilling a hole in the centre of the primary preform with the same diameter as the fugitive preform (Figure 3.6).

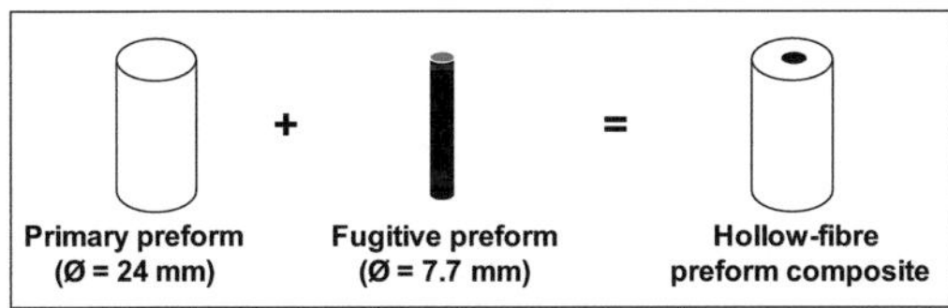

Figure 3.6: Preform composite assembly for hollow-fibres production.

3.5.3.2 Co-extrusion parameters

The co-extrusion experiments for hollow-fibre production were carried out using the capillary rheometer described before. The temperature was defined based on the characteristics of the thermoplastic binder material selected. A die with a half cone angle of 60°, 1 mm diameter and 16 mm length was used, so that the extrudate was 24 times smaller than the initial preform (R = 24:1). The speed of the piston was defined in view of the results attained to process the PZT solid-fibres.

3.5.4 Monofilament composite characterisation

During the co-extrusion experiments (solid and hollow-fibre production), lengths of 15 cm of each of the co-extruded materials were collected. The inner and the outer part area of monofilament composites were measured using the optical microscope equipped with a micrometer described before. The cross sections (perpendicular to the extrusion direction) of the monofilament composites were investigated by scanning electron microscopy (SEM; TS5136MM, Tescan) using a secondary electron detector. The samples were initially sputter coated with a Au-Pd layer (Cressington Sputter Coater 108auto). To avoid distortion of the ceramic-thermoplastic green bodies, the monofilament composites were fractured under liquid nitrogen.

3.6 Debinding and sintering

The green materials were prepared for debinding and sintering by placing them horizontally on ZrO_2 substrates. Each substrate could accommodate 18 fibres each of 9 cm length. The debinding schedules were based on the thermo-gravimetric analyses of the feedstocks. The binder removal of the extruded fibres was carried out at 1 °C/min up to 500 °C, holding for 2 h **[Hei05]**. With reference to the co-extruded monofilament composites, the organic vehicles were removed by slowly heating the materials (0.1 °C/min) up to 600 °C, holding for 10 hours.

All the debound materials were sintered by heating with 1 °C/min up to 1100 °C and holding for 2 h in a closed alumina crucible. To compensate the

PbO loss during sintering, a lead-rich powder ($PbZrO_3$ + ZrO_2; [PZZ]) was added to the alumina crucible. Since both an excess and a deficiency of PbO result in inferior piezoelectric properties, extruded and co-extruded solid fibres were sintered together, i.e., inside the same closed alumina crucibles containing Pb-enriched atmosphere (6 fibres of each condition was accommodated in each ZrO_2 substrate). An overview of the debinding and sintering steps is summarized in Table 3.5.

Table 3.5: Overview of the debinding and sintering steps

			Co-extruded fibres	
Step	**Parameters**	**Extruded fibres**	**CB as fugitive material**	**MCC as fugitive material**
Debinding	Heating rate	1 °C/min	0.1 °C/min	0.1 °C/min
	Final Temp.	500 °C	600 °C	600 °C
	Dwell time	2 h	10 h	10 h
	Atmosphere	air	O_2	air
	Furnace type	muffle kiln*	tubular**	tubular**
Sintering	Heating rate	1 °C/min	1 °C/min	1 °C/min
	Final Temp.	1100 °C	1100 °C	1100 °C
	Dwell time	2 h	2 h	2 h
	Atmosphere	PZZ	PZZ	PZZ
	Furnace type	muffle kiln*	muffle kiln*	muffle kiln*

* Pyrotech PY 12 H Muffelofen; ** STF 16/50, Carbolite Furnaces

3.6.1 Shrinkage

The theoretical shrinkage S_{theo} [%] was estimated using equation (3.2) **[Ger96]**.

$$S_{theo} = \left(1 - \sqrt[3]{\frac{\rho_G}{\rho_S}} \right) \times 100 \qquad (3.2)$$

where ρ_G is the initial fractional green density (calculated from the measured green density of the PZT powder, which depends on the powder loading of the feedstock) and ρ_s is the theoretical sintered density, assuming that it is equal to the initial powder density.

Due to the fact that the measurement of the linear shrinkage, i.e., measurement of green and sintered fibre diameter, was not feasible in the case of the co-extruded fibres, a calculated shrinkage is presented. The calculated shrinkage (S_{calc}[%]) was obtained applying equation (3.2), after measuring the porosity P_o [%] (equation (3.3)).

$$S_{calc} = \left(1 - \sqrt[3]{\frac{\rho_G}{\rho_{PZT} \times \left(\frac{100 - P_o}{100} \right)}} \right) \times 100 \qquad (3.3)$$

3.7 Microstructural analyses

The microstructure of the sintered solid-fibres obtained by conventional thermoplastic extrusion and by the co-extrusion process were analysed by means of porosity, grain size and phase analyses. The sintered hollow-fibres were not characterised within this work.

3.7.1 Porosity and grain size

3.7.1.1 Ceramographic preparation

The sintered fibres were cut into pieces of 10 mm length and horizontally embedded in a three-component epoxy cold-curing resin (Kulzer Technovit 4000, Heraeus Kulzer GmbH). For each processing condition, 6 pieces of fibres were embedded. The mounted specimens were subsequently grinded, lipped and polished as described in Table 3.6 **[Täf04]**. The equipment used was an automatic sample preparation system (Jean Wirtz Phoenix 4000).

Table 3.6: Ceramographic preparation steps

Process	Base	Abrasive	Lubricant	Speed [rpm]	Force [N]	Time [min]
Grinding	Abrasive Paper (Silicon Carbide # 500)		Water	150	Manually	Until flat surface
Lapping	Hard synthetic cloth [a]	6 μm diamond suspension [d]	Oil-based [f]	100	20	30
Polishing	Hard synthetic cloth [a]	3 μm diamond suspension [d]	Oil-based [f]	100	20	30
	Hard silk cloth [b]	1 μm diamond suspension [d]	Oil-based [f]	100	20	10
Fine polishing	Porous neoprene cloth [c]	Acid alumina suspension [e]	Water	100	15	30

[a] Texmet® P, Buehler GMBH

[b] DP-dur, Struers

[c] OP-chem, Struers

[d] DP-Suspension, P, Struers

[e] OP-AA Suspension, Struers

[f] DP-Lubricant Red, Struers

3.7.1.2 Etching

Samples were etched using an acid solution consisting of 95 mL distilled water, 4 mL hydrochloric acid (HCl 32%) and 1 mL hydrofluoric acid (HF 40%) to visualize the grain boundaries **[Täf04]**. The as-polished sections were exposed to the etching solution for about 5 seconds.

3.7.1.3 Microscopic examination

The microstructural details (as-polished: porosity; as-etched: grain size) were revealed in the SEM using a secondary electron detector. Samples were initially sputter coated by a Au-Pd layer to avoid electrical charging during examination. From each embedded fibre, 5 different micrographs were taken from the edge to the centre of the fibre. A magnification of 5 kX was applied for all as-polished samples.

3.7.1.4 Image analyses

The porosity was quantified using image analysis (Digital Micrograph 3.10.0, Gatan Inc.) by contrasting the pores (gray level = 0) and the material (gray level = 255). The grain size was determined through the linear-intercept method, using the software Lince (Version 2.3.2 beta), ensuring that an excess of 500 grains were sampled per material. This method determines mean grain size from the number of grains (intercepts) or grain boundaries (intersections) that intersect one or more lines of known length superimposed on a field of view or micrograph **[Chi02]**.

3.7.2 Phase analyses

The phase composition was analysed by X-ray diffraction (X'Pert PRO Multipurpose Diffractometer, PANalytical). Due to the low X-ray penetration depth into PZT (3 – 4 µm) **[Che92]**, it was possible to, at first, measure the as-sintered fibres in silica glass capillaries (Debye-Scherrer configuration) to reveal their surface phase composition. Afterwards, the fibres were crushed and the resulting powder mounted on a silicon single crystal substrate, in order to obtain information on the average phase composition of the fibre bulk material (Bragg-Brentano configuration). The performed angle scan was from 2Θ of 43° to 46° with a step width of 0.0167° within a scan time of 160 min (Cu K_α radiation, 40 mA heating current, 40 kV beam potential). The diffraction patterns were normalized with respect to the peak of highest intensity, since intensities depend on equipment and measuring conditions.

3.8 Electromechanical characterisation

For characterisation of the electromechanical properties, PZT fibres were cut into 3.5 mm long pieces and placed vertically into a PMMA sample holder (Figure 3.7) **[Bel09]**. Both fibre ends were coated with silver epoxy (Electrodag 5915, Acheson Colloids Company) to ensure electrical contact. The active free length of the fibres was thereby reduced to about 2.5 mm.

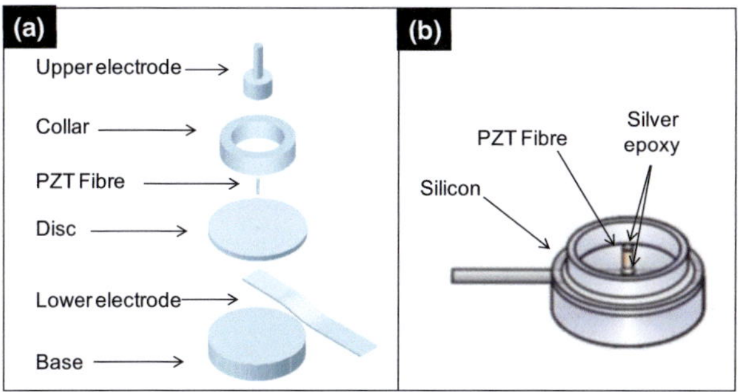

Figure 3.7: (a) Explored view of the sample holder **[Bel09]** and (b) set sample holder.

The fibres were immersed in silicon oil to avoid an electrical breakdown. Poling of the fibres was performed at room temperature applying an electric field of 3.5 kV/mm during 5 minutes. The fibres were subsequently measured in terms of their longitudinal free-strain as a function of the applied electrical field by using a waveform generator (Agilent 33120A) and a high-voltage amplifier (TreK Model 20/20C). With the use of a dynamic mechanical analyzer (DMA 7e, Perkin-Elmer), the free strain evolution was recorded for electrical fields ranging from -3 kV/mm to +3 kV/mm with a frequency of 2.778 mHz. This equipment applies forces to the sample with a sensitivity of 0.01 mN, thus enforcing nearly free-strain conditions. The detector, a linear variable differential transformer (LVDT), measures the vertical elongation of the sample with a sensitivity of 0.02 µm in the range from -5 mm to +5 mm. The measured displacement is recorded by the DMA software, divided by the active fibre length and plotted against the applied electrical field strength (butterfly loops). For each processing condition, 10 fibres were characterised.

4 Results and discussion – materials selection

The present chapter briefly discusses the criteria used for the selection of the primary (section 4.1) and fugitive materials (section 4.2) suitable for the development of feedstocks for the co-extrusion process of thin PZT fibres. Section 4.3 then describes the selection of the thermoplastic binder as the primary step in the development of the feedstocks to provide sufficient plasticity as well as proper rheological behaviour to the primary and fugitive materials during processing.

4.1 Selection of the primary material

Lead oxide, a major component of PZT, has a high vapour pressure above its melting point of 880 °C and is thus prone to volatilization at a typical sintering temperature of 1250 °C (**[Kin83a]**, **[Kin83b]**). One of the main technical challenges in the production of high quality PZT is maintaining compositional stoichiometry during sintering, since this strongly influences the electromechanical properties. Lead loss in small diameter fibres is a particular problem due to their high surface area/volume ratio **[Nel02]**. In view of that, with the aim of minimizing lead loss, the main requirement for the primary material selection was to determine a piezoelectric ceramic capable of sintering to full density at a low temperature.

In order to select a PZT powder fulfilling the aforementioned requirement, based on previous works in the literature (**[Hel99]**, **[Hei05]**, **[Hei07]**, **[Hei08]**, **[Hei09]**), two technologically important PZT-based powders were analysed: PZT-P505 and PZT-EC65 (Table 3.1). Following the steps for processing thin PZT solid-fibres by the conventional thermoplastic extrusion **[Hei05]**, the different PZT powders were initially coated with a surfactant. These results will be discussed in the section 4.1.1. Once a suitable concentration of surfactant was chosen, feedstocks containing 58 vol. % of PZT were prepared by using low density polyethylene as the thermoplastic binder **[Hei05]**. This solid loading was selected because it should provide a viscosity within a reasonable working range for the extrusion process as well as a sufficient sintered density **[Hei08]**. Finally, the effect of varying the sintering temperature between 1050 and 1200 $^\circ$C in steps of 50 $^\circ$C was investigated in the thin extruded PZT-based fibres and their microstructure and electromechanical properties characterized. These results will be discussed in the section 4.1.2.

4.1.1 PZT-based powder: surfactant effect

In the PZT-thermoplastic binder systems, the surface of the PZT powder is hydrophilic, while the binders to be investigated in this study (Table 3.3) possess a rather hydrophobic character. When the ceramic-polymer systems are mixed, the PZT particles tend to agglomerate due to attractive van der Waals (vdW) forces, resulting in difficulties to obtain a homogeneous distribution of PZT particles in the polymer. To overcome the vdW attraction forces and stabilize these mixtures against agglomeration, stearic acid was utilized to induce steric repulsion of the polymer chains adsorbed onto the ceramic particle surfaces. The selection of this surfactant was based on the work of McNulty et al. **[Mcn99]**, in which was shown that, stearic acid is a carboxylic acid with a large tendency to bond to the PZT powder surface. The degree of dispersion of the powder in the molten thermoplastic binder has a remarkable effect in obtaining homogeneous feedstocks and, as a result, defect-free sintered bodies. Additionally, the better the powder is dispersed in the polymer medium, the lower the resultant viscosity; and by extension, the higher the usable solids loading for a given viscosity range.

One important requirement that a surfactant should satisfy is that it should not alter the mechanical, thermal, or rheological properties of the thermoplastic binder once the ceramic powder is mixed. This means that the surfactant which coats the powder should remain on the powder, and any excess surfactant remaining from the coating process must be removed in order to eliminate the possibility of incorporating the excess surfactant into the polymer phase **[Mcn99]**. In view of that and in order to define the suitable concentration of stearic acid to coat the as-received PZT powders based on equation (3.1), increasing amounts (layers) of stearic acid were analysed. The particle size distribution of the coated powder was then measured. These results are presented in Table 4.1 for the PZT-P505 (batch 1). For the case of the PZT-EC65, the suitable concentration was chosen based on the work of Heiber et al. **[Hei05]**, where the same PZT powder was used.

Tale 4.1: Particle size distribution (PSD) of the PZT-P505 (batch 1) powder uncoated and coated with different layers of stearic acid

PSD [µm]	Uncoated	1-layer	2-layers	3-layers	4-layers
d_{10}	1.35	0.65	0.60	0.57	0.57
d_{50}	2.66	0.98	0.89	0.85	0.87
d_{90}	4.28	1.98	1.92	2.07	1.97

Referring to Table 4.1, a decrease in PSD is observed in the presence of the stearic acid (uncoated compared with the different stearic acid layers), emphasizing the effectiveness of the surfactant used. However, increasing the concentration of stearic acid beyond 3-layers, did not result in a decrease in PSD. Hence, it was assumed that 3-layers of stearic acid were sufficient to completely coat the PZT powder. The use of higher concentrations would probably affect the thermoplastic binder phase once the coated powder is mixed for feedstock preparation.The next logical step in the process was to test the PZT powder coated with 3-layers to examine if there is a marked improvement over uncoated powder in terms of mixing. Figure 4.1 represents the mixing curve of PZT feedstocks (52 vol. % PZT + LDPE as the thermoplastic binder) with and without the use of stearic acid (feedstocks n° 8 and 9, Table 3.4). As expected, the difference in torque was significant. Similar mixture using uncoated powder and a PZT loading of 58 vol. % was performed, but difficulties in thoroughly dispersing the powder in the thermoplastic binder led to immeasurably high torques.

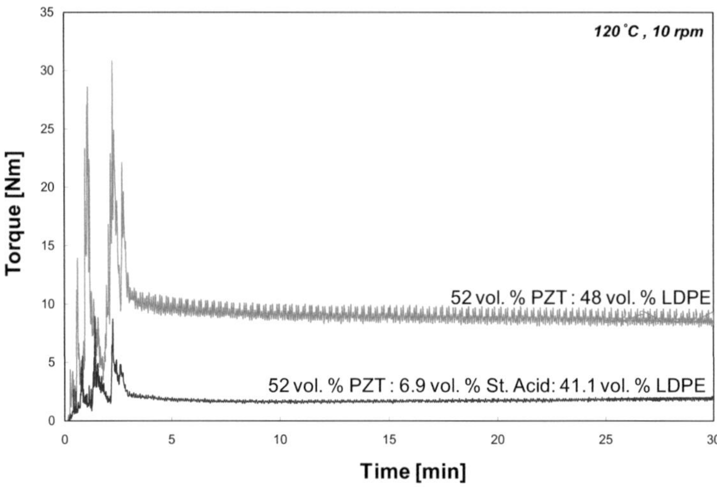

Figure 4.1: Mixing curves (torque as a function of time) of PZT feedstocks with 3-layers and without the use of stearic acid.

4.1.2 PZT-based powder: sintering behaviour

The feedstocks used for the extrusion of the PZT solid-fibres were the feedstocks n° 11 (PZT-P505) and 13 (PZT-EC65) listed in Table 3.4. The processed fibres were sintered in a lead-rich atmosphere, as discussed in Chapter 3. Figure 4.2 presents the dependence of strain on electric field for the thin extruded fibres ($\varnothing \sim 250$ µm) processed using the different PZT-based powders and sintered at different temperatures.

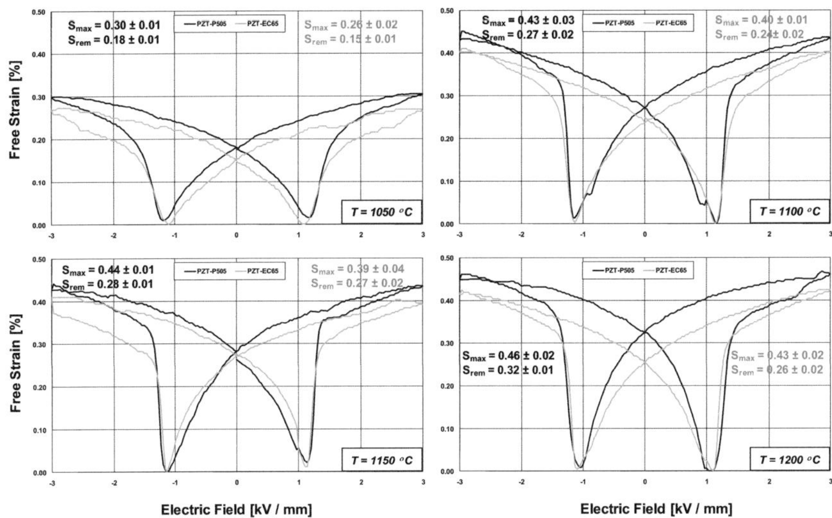

Figure 4.2: Free strain as a function of electric field for PZT-P505 (58 vol. % PZT (1) + LDPE feedstock) and PZT-EC65 (58 vol. % PZT (3) + LDPE feedstock) extruded PZT fibres sintered at different temperatures. The butterfly curves represent the average values recorded for five fibres of each condition. The average maximum free strain (S_{max}, at 3 kV/mm) and remnant strain (S_{rem}) is additionally represented for each temperature (left side PZT-P505; right side PZT-EC65).

Figure 4.2 indicates that the increase in sintering temperature was followed by an improved electromechanical performance (S_{max} and S_{rem}), independent of the used PZT powder. However, in general, PZT-P505 shows a 10% higher maximum and remnant strain compared to PZT-EC65, independent of the sintering temperature. In view of these results, the PZT-P505 powder was selected within this work as the primary material. This powder is a PZT ceramic substituted with the lead-free complex compound $Sr(K_{0.25}Nb_{0.75})O_3$ (SKN). It has a morphotropic composition with excellent sensor as well as actuator properties. Materials of this compositional series

61

are characterised by a relatively low coercive field and high piezoelectric activity (**[Hel99]**, **[Hel08]**).

Moreover, it is worth to mention that the feedstocks containing PZT-P505 required lower torques during mixing than the feedstocks containing PZT-EC65. This is in addition an important parameter to be considered for the co-extrusion process, since as high as the torque measured during mixing, as high as the pressure necessary to co-extrude. Another advantage of the PZT-P505 powder is their finer particles when compared to PZT-EC65 (Table 3.2), since the finest feature that can be micro-fabricated is limited by the granularity of the extrusion body itself, which in turn is proportional to the particle size of the powder.

In order to select a suitable sintering temperature to be further applied for the co-extruded materials, the microstructure (porosity and grain size) of the sintered PZT-P505 fibres was analysed. Figure 4.3 correlates the microstructure characteristics with the electromechanical performance of PZT-P505 based fibres sintered at different temperatures.

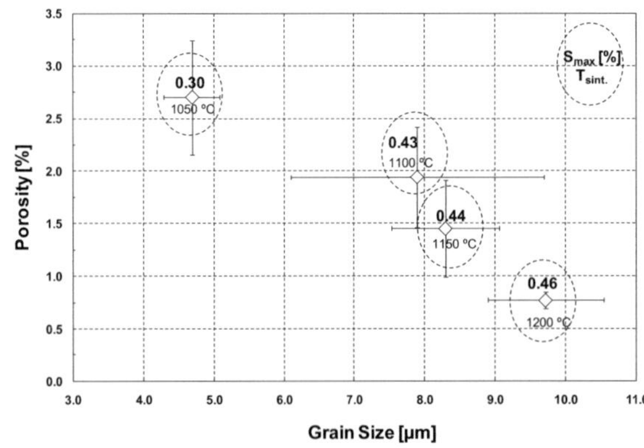

Figure 4.3: Microstructure characteristics (porosity and grain size) correlated with the maximum strain (S_{max}, at 3 kV/mm) for PZT-P505 fibres sintered at different temperatures (T_{sint}).

From Figure 4.3 it is obvious that the average grain size increased with sintering temperature; especially above 1050 °C a marked increase took place. With increasing temperature, further grain growth occurs. Conversely, as expected, the porosity decreased as temperature increased, achieving an almost complete densification at higher temperatures. In terms of maximum attained strain, the best electromechanical activity was verified for the fibres sintered at 1200 °C. This statement is in good agreement with the work of Helke et al. **[Hel99]**, in which for the SKN co-doped morphotropic PZT, an increase in field induced strain with grain size was measured. A reasonable explanation for this phenomenon might be the phase assemblage which changes with grain size; however, differences in the behaviour of field induced strain to changes in grain size may be influenced by a large numbers of factors **[Kun10]**. However, as this topic is out of the scope of the present work, further analysis and correlations between grain size and strain is suggested in order to fully comprehend the system studied here. As pointed out, the aim is to select a low-temperature sinterable PZT powder. Based on this requirement, the sintering temperature selected for further investigations was 1100 °C, since the microstructure and electromechanical performance of those fibres appeared to be reasonable good.

4.2 Selection of the fugitive material

The requirements for the selection of the fugitive material included: (1) determining a substance that could be completely removed after the micro-fabrication, including materials which can be dissolved, evaporated, or removed by reaction, and (2) the material should reveal an apparent viscosity in the range of the PZT feedstock (58 vol. % PZT), which excludes the possibility of working with a pure thermoplastic material as the fugitive substance. The fugitive material should be then able to be mixed with a thermoplastic material (surface compatible). Due to the fact that carbon black is inexpensive and typically formulated to enhance dispersion in polymers, it has been extensively used as the fugitive material in ceramic co-extrusion processes (**[Kov97]**, **[Van98]**, **[Cru98]**, **[Buc01]**, **[Koh02]**, **[Kim04]**, **[Yoo05]**, **[Cru07]**, **[Dro09]**). In view of that, in the first step of the present study, carbon black (CB) was selected as fugitive filler. By investigating the degradation behaviour through thermal analyses of different commercially available CBs (Figure 4.4), BP$^{®}$ 120 (Table 3.1) was selected within this work. However, despite that BP$^{®}$ 120 carbon black presented the lowest decomposition temperature among the investigated ones (664 °C), this temperature is indeed too high, since PZT starts sintering at 600 °C **[Ham98]**. It is known that the high temperature decomposition of the carbon can lead to a local reduction in the partial oxygen pressure, increasing the rate of generation of unstable and dissociative lead sub-oxides, thereby raising the amount of lead evaporation **[Kah92]**, which might significantly affect the final electromechanical performance of the PZT materials. Therefore, in the second step of the present work, microcrystalline cellulose was selected to be used as an alternative to the carbon black. The selection was done based on the fact that the MCC material fulfilled the aforementioned requirements for the fugitive substance selection: the burnout is complete around 500 °C (Figure 4.4) and when mixed with a thermoplastic binder and a surfactant, it reveals a viscosity in the range of a PZT-based feedstock, depending on the MCC loading (Figure 4.5). MCC is used as additive to promote green strength of clays, as reinforcement for thermoplastic composites and in addition as gelling agent in the food industry (**[Bat75]**, **[Pal07]**).

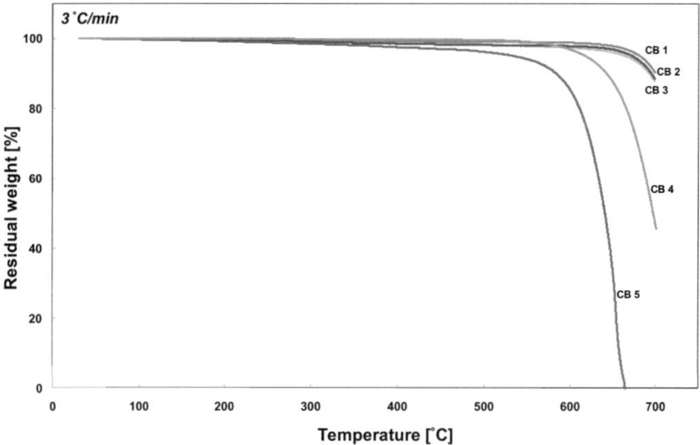

Figure 4.4: Decomposition profiles (in air) of the investigated commercially available carbon blacks (CB 1 to CB 5) and of the microcrystalline cellulose powder (MCC). *Note: CB 5 corresponds to the BP® 120.*

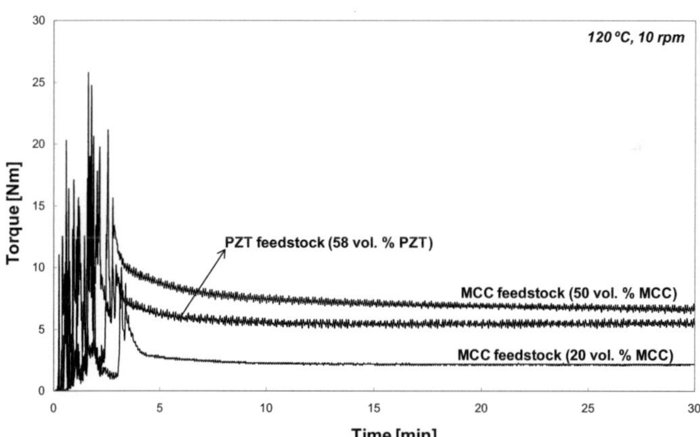

Figure 4.5: Mixing curve (torque as a function of time) for the primary feedstock (58 vol. % PZT) and for different solid loadings (20 and 50 vol. %) of MCC based feedstocks.

4.3 Selection of the thermoplastic binder material

In view of the considerations presented in Chapter 2, the first step in the development of the thermoplastic co-extrusion process is the selection of the optimal organic binder additives. However, choosing an appropriate binder is usually a compromise. Some pre-requisites must be taken into account before the selection. An extensive literature on polymers as binders for ceramics is available (**[Edi86]**, **[Ree95]**), although most of it concerns the binder removal from the green body (**[Lew97]**, **[Hrd98]**, **[Kna06]**). The foremost criterion for binder selection is that it should provide suitable rheological properties to the ceramic feedstock during processing. How do the polymer properties such as transition temperatures influence the processing behaviour of the ceramic feedstocks? How do the interactions between the thermoplastic binders and the ceramic particles influence the processability of the feedstocks? The influence of the binder in the processability of the feedstocks might be predicted by a systematic characterisation of the thermoplastic binders and their corresponding filled system. The objective of the present section is to understand and discuss the effects of four different thermoplastic binder characteristics on the processing behaviour of PZT feedstocks (PZT-P505 based feedstocks, once selected as the primary material), in order to select the most suitable for the co-extrusion of PZT fibres. Obviously, a large number of polymers that could be used for the thermoplastic shaping process of ceramics is available. However, the binders selected to be investigated (Table 3.3) have been extensively used in the thermoplastic extrusion field. Additionally, those binders are commercially available at an adequate price. Polystyrene (PS) has been applied as the organic vehicle for the preparation of extruded zirconia plates **[Pay98]**. Low density polyethylene (LDPE), a polyolefin containing a highly branched chain structure **[Bry99]**, has been successfully used for the extrusion of polycrystalline ceramic fibres, such as PZT, silicon carbide, and lanthanum strontium manganate (**[Weg98]**, **[Hei05]**, **[Cle07]**). The poly(ethylene-co-ethyl acrylate) (EEA) copolymer has been mixed with conductive fillers to form electrically conductive polymer composites **[Fel04]**. The poly(isobutyl methacrylate) (PiBMA) resin was selected with the function to blend the EEA binder. The reason for blending the EEA is based on the fact that its high flexibility can be modified by incorporating a brittle thermoplastic polymer **[Koh04]**. This blended system has been widely used for the thermoplastic processing of ceramics by co-extrusion (**[Cru98]**, **[Koh02]**, **[Yoo05]**, **[Cru07]**).

In the course of this section, initially, the unfilled thermoplastic binders are characterised. Following, the rheology and processability of PZT-filled and unfilled thermoplastic binders are compared and discussed. Finally, using the different binders, PZT solid thin fibres are processed by thermoplastic extrusion to analyse the die swell.

4.3.1 Unfilled thermoplastic binders: thermal behaviour analyses

Figure 4.6 represents the thermograms of unfilled thermoplastic binders, and Table 4.2 lists the information derived from these measurements. With reference to the amorphous polymers, the values of the T_g indicated in Table 4.2 correspond to the average between the onset and the final temperatures of the inflection from the baseline in Figure 4.6. Hence, the shift from the baseline verified between 106 and 111 °C for the PS, and the inflection between 52 and 70 °C for the PiBMA, correspond to the ranges of their T_g. For the LDPE and EEA copolymer, the peak temperature occurring at 107 and 104 °C is assigned to their T_m, respectively. The studied polymer blend (EEA + PiBMA) is determined to be compatible because the transition temperatures are slightly shifted when compared with the corresponding pure components. An incompatible blend would only show the characteristic DSC responses of the individual components **[Sim03]**. The importance in determining the transition temperatures of the organic binders relies on the set-up of the processing temperatures.

The mixing and co-extrusion step should be performed above the T_m and the T_g values for semi-crystalline and amorphous polymers, respectively, in order to facilitate the wetting of the powder.

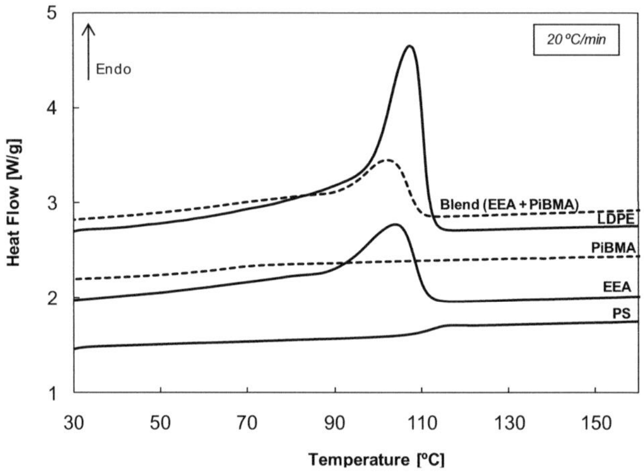

Figure 4.6: DSC thermograms of the unfilled thermoplastic binders.

The TGA of the organic systems are shown in Figure 4.7. The temperature at which weight loss onset occurs (T_{onset}), the temperature that corresponds to the maximum decomposition rate (T_{max}) and the temperature for 50% weight loss (T_{50}), was derived from these measurements and listed in Table 4.2. Additionally, Table 4.2 represents the parameters for the pure stearic acid and for the PZT powder coated with 3-layers of stearic acid (coated PZT, PZT-P505 batch 1). The parameters of the coated PZT were determined after normalizing the TGA results for the amount of stearic acid.

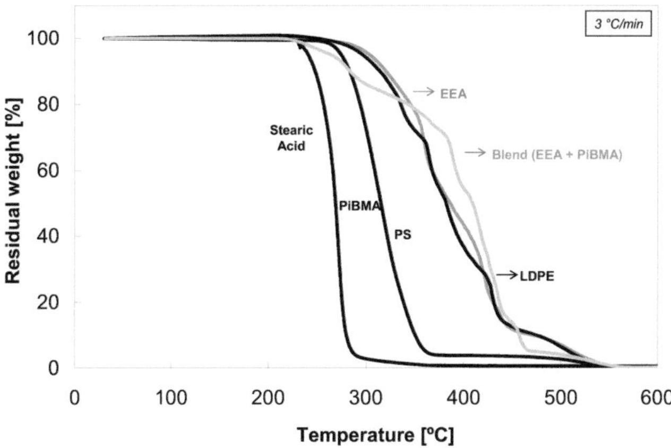

Figure 4.7: Decomposition profiles of the unfilled organic binders investigated. *Note: The profile of the stearic acid (surfactant) is additionally represented.*

Table 4.2: Information derived from the DSC and TGA measurements of the organic materials

Organic material	T_m [°C]	T_g [°C]	T_{onset} [°C]	T_{50} [°C]	T_{max} [°C]
PS	amorphous	111	260	315	315
PiBMA	amorphous	61	226	269	271
LDPE	107	-	275	382	363
EEA	104	-	270	386	359
Blend	102	68	230	409	387
St. Acid	62	-	150	215	223
Coated PZT*	-	-	220	275	280

* PZT-P505 batch 1

Considerable differences in the thermal decomposition of the thermoplastic binders are observed in terms of both degradation temperature (Table 4.2) and pathway (Figure 4.7). While the decomposition of the amorphous polymers (PiBMA and PS) occurred in a continuous stage, the semi-crystalline binders (EEA and LDPE) showed distinct degradation steps.

The addition of the amorphous PiBMA to the EEA polymer led to a decreased T_{onset} of the blend when compared to the pure EEA, however, T_{50} and T_{max} increased. Additionally, as for semi-crystalline binders, the blend shows distinct degradation steps. The amorphous binders presented a lower final decomposition temperature than the semi-crystalline thermoplastics.

The study of the thermal performance of the organic compounds stems from the need to correlate their degradation behaviour to the processing steps. Processing (feedstock preparation, preform fabrication and co-extrusion) should be carried out at temperatures significantly below the binder T_{onset} to avoid drastic changes in the flow behaviour of the feedstocks due to degradation of the thermoplastic material. In fact, for the case of the thermoplastic binders investigated within this study, the limiting temperature for processing is the T_{onset} of the coated PZT (lowest T_{onset} when compared with the thermoplastic binders). The retention of the stearic acid during processing is necessary for the consistency of ceramic powder volume loading. In addition, the degradation of the stearic acid during processing would change the flow behaviour of the composition. It is noteworthy that the pure stearic acid starts decomposing at 150 °C, but an increase of 70 °C on the T_{onset} is measured when this molecule is adsorbed onto the particle surface of the PZT. This is attributed to the strong physical bond between the stearic acid head group (COOH, Table 3.3) and the ceramic surface **[Mcn99]**.

TGA is important in identifying temperature regions of rapid mass loss (T_{max}), which can be used as an initial guide in the set-up of the debinding process of the as-shaped ceramics. The decomposition reactions must be fast enough as to provide the total burnout of the organic components during the heating cycle within a reasonable timescale, but must be slow enough as to avoid sudden formation of volatile gas products, which may rise to bubble and crack formation. As previously discussed, the binder should be completely removed before sintering of the ceramic starts. It was verified that no residue was left for any of the investigated binders at 600 °C (Figure 4.7), the temperature at which PZT starts sintering in case of PbO excess **[Ham98]**.

4.3.2 Unfilled and filled thermoplastic binders: rheology and processability

It is well known that the polymers provide the rheological properties required for ceramic feedstocks. However, the interactions between the thermoplastic binders and the ceramic particles might additionally affect the flow behaviour, and, as a result, the processability of the feedstocks is influenced. In this context, this section discusses the rheological behaviour of the unfilled polymers in comparison with the corresponding filled mixtures.

Table 4.3 indicates the temperatures at which the mixing (feedstock preparation) and the rheological investigations were carried out. The formulation of these feedstocks is presented in Table 3.4. The temperatures for the mixing step and for the rheological investigations were accurately controlled, i.e., the temperatures indicated in Table 4.3 corresponds to the temperatures measured inside the mixing chamber. The displayed mixing temperature corresponds to the second mixing step (as described in Chapter 3). The mixing and the first temperature for the rheological investigations (T1) were set as low as possible to suppress thermal degradation of the organic materials (T_m, T_g < T1 < T_{onset}) and T4 was set to be lower than the T_{onset} of the thermoplastic binder (T4 < T_{onset}). For the case of the PZT feedstocks, the T_{onset} of the surface-modified PZT (coated PZT) was the limiting working temperature (Table 4.2). For the polystyrene-containing systems, elevated temperatures (T1 >> T_g) were required due to the high torque values observed for the unfilled PS, an indicative of a highly viscous binder. Besides the equipment imposes a limit for maximum torque reading, high torque values may result in an elevated applied pressure during co-extrusion. In addition, Table 4.3 gives the maximum and the equilibrium torque measured at the second mixing step (for a better interpretation of Table 4.3, refer to Figure 2.10). The equilibrium torque represented is the average value for the last five minutes of steady-state. Both parameters are dependent on the melt viscosity of the thermoplastic binder. It can be seen that LDPE reveals the lowest viscosity (lowest torque), whereas PS shows the highest one (highest torque). The addition of PiBMA to the EEA co-polymer did not significantly influence its behaviour. For the unfilled polymers, the maximum torque corresponds to the heat transfer necessary to completely melt the thermoplastic material. For the case of the filled polymers, the maximum torque value is an indicative of the energy necessary to fill voids between particles and cover their surface.

Table 4.3: Mixing parameters and temperatures used for the rheological investigations for the unfilled polymers and their feedstocks

Systems	Mixing (at 10 rpm)			Rheological characterizations Temperature, T [°C]			
	T [°C]	Maximum torque [Nm]	Equilibrium torque [Nm]	T1	T2	T3	T4
Unfilled PS	150	43.10	15.89 ± 0.20	150	160	170	180
Unfilled LDPE	120	8.90	3.30 ± 0.01	120	130	140	150
Unfilled EEA	120	17.14	7.11 ± 0.08	120	130	140	150
Unfilled Blend	120	17.00	6.52 ± 0.21	120	130	140	150
58 vol. % PZT + PS	180	53.08	11.00 ± 0.43	150	160	170	180
58 vol. % PZT (1) + LDPE	120	20.04	8.75 ± 0.21	120	130	140	150
58 vol. % PZT + EEA	120	37.24	13.00 ± 0.30	120	130	140	150
58 vol. % PZT + Blend	120	38.14	13.31 ± 0.14	120	130	140	150

Besides determining the aforementioned parameters during mixing, the torque-rheometer is, in addition, useful in providing information on the flow behaviour of the materials. This can be attained when considering the torque-rheometer measuring head acting as two adjacent concentric-cylinder rheometers (**[Bly67]**, **[Goo67]**). Neglecting end effects, the following equation is applied in concentric-cylinder rotational rheometers **[Mac94]**:

$$\eta_{app} = \frac{M}{\Omega} k \tag{4.1}$$

where η_{app} is the apparent viscosity, M the torque moment, Ω the angular velocity and k is a constant which depends on the dimensions of the rheometer (internal and external radius, R_i and R_e, respectively, and cylinder length L).

A schematic of a torque- and a concentric-cylinder- rheometer and the corresponding analogy are given in Figure 4.8. In view of this representation and considering equation (4.1), the torque moment (M) can be related to the imposed shear stress (τ) and the rotor speed (N) to the shear rate ($\dot{\gamma}$):

$$\tau = k_1 M \tag{4.2}$$

$$\dot{\gamma} = k_2 N \tag{4.3}$$

therefore,

$$\eta_{app} = \frac{\tau}{\dot{\gamma}} = \frac{k_1 M}{k_2 N} \tag{4.4}$$

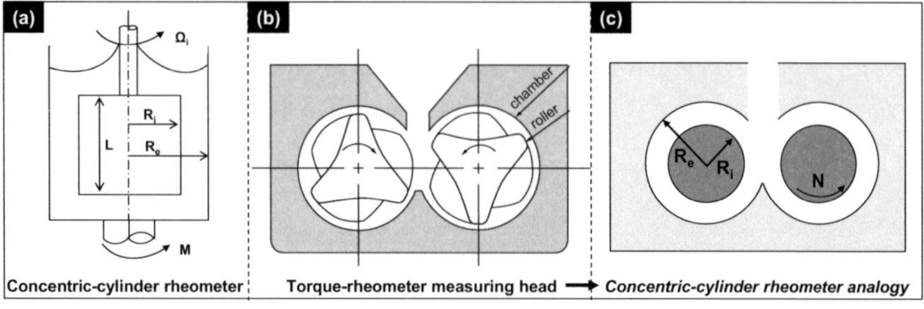

Figure 4.8: Schematic of (a) concentric-cylinder and (b) torque- rheometer, and (c) the torque-rheometer analogy to a concentric-cylinder rheometer.

The melt viscosity of a polymer depends on the temperature used, which in turn has a pronounced effect on the processing step. The temperature-viscosity relationship can be defined using an Arrhenius type equation **[She99]**:

$$\eta_{app} = \frac{\tau}{\dot{\gamma}} = Ae^{\frac{E_f}{\overline{R}T}} \tag{4.5}$$

where η_{app} is the apparent viscosity at temperature $T\,(K)$, A is the frequency term depending on the entropy of activation for flow, E_f is the energy of activation for viscous flow and $\overline{R}$ is the universal gas constant (8.314 J/molK). Combining (4.4) with (4.5), derived from the rheological characterisations performed using torque-rheometry, Figure 4.9 represents Arrhenius-type plots of torque/rpm-temperature data for the unfilled polymers and their corresponding feedstocks.

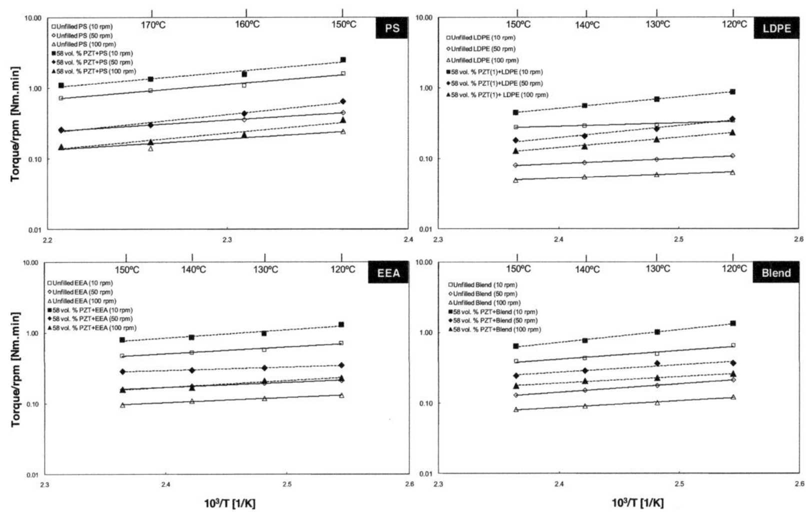

Figure 4.9: Temperature dependence of viscosity (torque/rpm) for the unfilled polymers and their corresponding feedstocks. *Note: temperature axes are different.*

Referring to Figure 4.9, with respect to the temperature effect, all unfilled thermoplastic binders and feedstocks obeyed an Arrhenius relation. Additionally, as expected, Figure 4.9 points out that all measured systems exhibit pseudoplastic behaviour for the temperature and shear ranges analysed, because the apparent viscosity (torque/rpm) decreased with increasing shear rate (rpm). Polymers under shear tend to align their chains. At low shear rates, Brownian motion of the segments occurs, so that polymers can coil up at a faster rate than they are oriented and become to some extent disentangled. At high shear rates, such re-entangling rates are slower than the orientation rates, and hence the polymer is apparently less viscous **[She99]**. Filling the polymers increases the resistance to flow. The addition of ceramic particles to the system, besides decreasing the effective volume of polymer, leads to a reduction in the polymer chain mobility.

The usefulness of Arrhenius representations is related to the potential correlation between the feedstock preparation step and the co-extrusion step: by determining the torque at a given temperature during compounding, the pressure necessary to co-extrude can be estimated. For example, in this study, equilibrium torque values down to 3 Nm at 10 rpm were an indication of very low viscous compositions, allowing the material to flow by gravity during the extrusion process when using a capillary die with 1-mm diameter. On the other hand, for feedstocks with equilibrium torque values higher than 30 Nm at 10 rpm, the pressure during extrusion through a 1-mm diameter die increased dramatically. The aforementioned correlation is of great importance for the co-extrusion technique; mixing and extrusion should be carried out at the same temperature (defined during the rheological characterisations) due to the sensitivity of viscosity to the temperature changes that polymers display.

Moreover, from Arrhenius plots it is feasible to determine the activation energy of viscous flow (E_f), a measure of the resistance to flow of a given material. During the flow of any liquid, layers of molecules slide over each other, where intermolecular forces oppose the motion and result in an energy barrier for viscous flow. Figure 4.10, derived from Figure 4.9, shows the activation energy of viscous flow for the unfilled polymers and their corresponding feedstocks. The results found for the unfilled polymers are in good agreement with values for E_f previously reported in the literature (**[Bly67]**, **[Goo67]**). When comparing the unfilled polymers with their corresponding feedstocks, an increase of E_f with the presence of the ceramic particles is observed for the PS and LDPE feedstocks, because, as previously discussed, fillers lead to a decrease in the polymer chain mobility. However, for the case of the EEA and the blend, this assertion is not applicable. One comprehensible explanation can be the presence of the ester functional group($-RCOOR'-$) in the EEA and PiBMA binders (Table 3.3). This fact confers a polar character to the aforementioned binders, allowing them to perform as dispersants and enabling a better wetting of the inorganic filler.

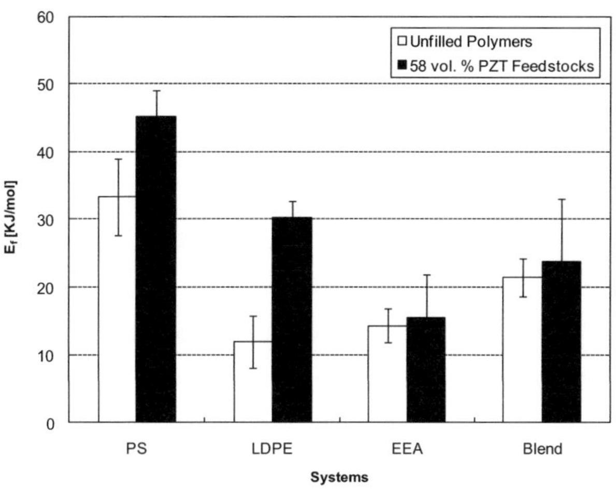

Figure 4.10: Activation energy of viscous flow for the unfilled polymers and their corresponding feedstocks at rotor speeds ranging from 10 to 100 rpm.

It is considered that the energy barrier for viscous flow reflects the contributions of polymer-particle and particle-particle interactions. For all PZT feedstocks, the particle-particle interaction should be similar due to equal conditions of PZT powder and surfactant. Therefore, it is reasonable to correlate the E_f values of the feedstocks with the polymer-particle interaction; for a comprehensive comparison, the E_f values of the unfilled polymers should additionally be taken into account. Thus, referring to Figure 4.10, a strong interaction of LDPE chains with PZT particles is assumed due to the abrupt change in the flow activation energy observed from the unfilled polymer and its feedstock. The 58 vol. % PZT + PS compound also displayed a high E_f; however, so did the unfilled PS. This can be explained by the low MFI of the referred binder (Table 3.3), what is an indicative of a high molecular weight polymer. The higher the molecular weight of a thermoplastic material, the greater the chain entanglements, which results in an increase of the melt viscosity, an occurrence that decreases processing ease.

The feedstocks containing the EEA and the EEA blend did not show a strong polymer-particle interaction, as observed on the similar values of E_f for the unfilled polymers and their feedstocks. A strong polymer-particle interaction is desired to avoid phase separation during extrusion, a phenomenon whereby the effective solid loading of the system is increased, resulting in altered flow properties. Taking into account the huge standard deviation verified for these feedstocks, it is reasonable to assume that at a particular shear rate, the PZT feedstock presented a lower E_f than its corresponding unfilled polymer. Despite the presence of the ester functional group formerly mentioned, this might be an indication of the degradation of the thermoplastic binder in the presence of the ceramic powder. It is important to mention that after compounding those feedstocks, a colour change was observed, an indication of the degradation phenomenon (results not shown). High shear stresses in the particle-crowded condition may break the polymer into lower molecular weight fragments **[Wig01]**. The decrease in the molecular weight of the binders diminishes the activation energy of viscous flow. However, further investigations should be carried out to confirm the occurrence of this phenomenon.

Assuming that E_f reflects the contributions of polymer-particle and particle-particle interactions, the PZT/LDPE feedstock shows the strongest polymer-particle interaction. This might be mainly attributed due to its greatly branched chain structure, which increases the probability to form multi-adsorption sites on the ceramic particles.

4.3.3 Fibres extrusion

The final process step for the selection of the thermoplastic binder material was to extrude the PZT feedstocks through a 300 µm diameter die and measure their die swell. Due to the drawbacks of the polystyrene feedstock discussed to this point, such as its high apparent viscosity values (Figure 4.9), this composition was not extruded.

Table 4.4 points out the extrudate diameter and the die swell (S_w) values for the PZT extruded fibres.

Table 4.4: Extrudate diameter and die swell of the fibres extruded through a 300 µm diameter capillary at 120 °C

Feedstocks	Extrudate diameter [µm]	Die swell S_w [%]
58 vol. % PZT (1) + LDPE	321 ± 5	7
58 vol. % PZT + EEA	359 ± 6	20
58 vol. % PZT + Blend	330 ± 4	10

Table 4.4 indicates that the 58 vol. % PZT (1) + LDPE feedstock presents the lowest die swell value, whereas the 58 vol. % PZT + EEA showed the highest swelling ratio. The addition of PiBMA to the EEA co-polymer (blend feedstock) decreased the S_w result of the EEA feedstock. The die swell effect is an important parameter to be taken into account for composites shaped by co-extrusion. Besides the fact that feedstocks presenting high die swell may generate stress between the different co-extruded materials, this parameter is additionally essential for achieving final structures with geometries identical to the original preform.

In view of the results attained, based on the fact that the selection of the ideal binder for co-extrusion requires consideration of parameters such as degradation during compounding, flow behaviour as well as the final product dimension desired, the low density polyethylene was selected within this work as the thermoplastic binder material for the co-extrusion of thin PZT-based fibres. Under the investigated processing conditions, this binder exhibits low viscosity with a high solid loading, no indication of degradation, a strong particle-polymer interaction as well as a comparatively low die swell.

4.4 Summary

The objective of the work in this chapter was to select the primary, fugitive and thermoplastic binder materials for the preparation of feedstocks for the co-extrusion of PZT thin fibres. The main results and conclusions are as follows:

For the primary material, two technologically important PZT-based powders were characterised: PZT-P505 and PZT-EC65. Due to the high ratio to surface area that fibres present, in order to minimize lead loss during sintering, the aim was to select a low-temperature sinterable PZT powder with suitable microstructure and electromechanical properties. The first step was to determine the proper amount of stearic acid (surfactant) to coat the PZT powders, and thus, stabilize the mixtures against agglomeration. For the case of PZT-EC65, the suitable concentration was chosen based on previous works on the literature. For PZT-P505, by testing increasing concentrations of stearic acid and measuring the coated particle size distribution, it was found that 3-layers of stearic acid were sufficient to completely coat the powder. PZT feedstocks (58 vol. %) were then prepared and extruded to produce 250 µm diameter sintered fibres. Their electromechanical performance as a function of sintering temperature (1050, 1100, 1150 and 1200 $^{\circ}$C) was analysed and compared. In general, it was verified that the PZT-P505 powder reveals a 10% higher maximum and remnant strain compared to PZT-EC65, independent of the sintering temperature. Additionally, PZT-P505 developed a reasonable good microstructure (porosity and grain size) and electromechanical performance even when sintered at 1100 $^{\circ}$C, and was therefore selected for this work as primary material for the micro-fabrication of PZT fibres by co-extrusion.

In view of the fugitive material selection, two substances were considered to be investigated within this work: carbon black and microcrystalline cellulose. For the first part of this study, carbon black (CB) was selected based on the fact that it has been extensively used as the fugitive material in ceramic co-extrusion processes. A suitable CB powder for the present work was selected studying the degradation behaviour of different commercially available CB powders. For the second part, microcrystalline cellulose was selected as an alternative to carbon black. The selection was done based on the fact that the MCC material experiences complete burn out before sintering of PZT takes place. Additionally, with the use of stearic acid in the composition, this substance is able to be mixed with a thermoplastic binder, achieving an apparent viscosity in the range of the primary feedstock (58 vol. % PZT).

Concerning the thermoplastic binder selection, the effects of four different thermoplastic binder systems (polystyrene, low density polyethylene, poly(ethylene-co-ethyl acrylate) and a blend of poly(ethylene-co-ethyl acrylate) and poly(isobutyl methacrylate)) on the processing behaviour of 58 vol. % PZT containing feedstocks were analysed and discussed. The thermal behaviour of the unfilled binders was characterised using differential scanning calorimetry and thermo-gravimetric analyses. The results attained were essential in determining the optimum temperature range for processing (feedstocks preparation, preform fabrication and co-extrusion). The rheological behaviour of the unfilled polymers in comparison with the corresponding ceramic-filled mixtures was analysed by means of torque rheometry. Using an Arrhenius-type approach, the temperature dependency of viscosity was determined and the energy of activation for viscous flow calculated. The Arrhenius representations were shown to be useful in estimating the co-extrusion behaviours of the compounds at a given temperature. The fact that the different thermoplastic binders indicate different interactions with the PZT ceramic resulted in dissimilarities in the activation energies for flow. The die swell effect was investigated on 300 µm diameter PZT fibres obtained by conventional thermoplastic extrusion technique. Finally, low density polyethylene was selected within this work as the thermoplastic binder material due to its relatively low viscosity with a high solid loading, no indication of degradation, strong particle-polymer interaction as well as the comparatively low die swell.

5 Results and discussion- co-extrusion of PZT fibres: processing

After the selection of the primary, fugitive and thermoplastic binder materials (Chapter 4), the suitable parameters for the solid and hollow-fibres co-extrusion were determined. This includes the selection of the solid loading and appropriate conditions during extrusion (shear rate and viscosity). In view of that, initially, this chapter deals with a rheological characterisation of the feedstocks. Following, it reports how the interface shape in the extrudate might be affected by the difference in the rheological properties of the co-extruded materials. The binder and fugitive material removal of the different geometries (solid and hollow) processed fibres is then presented.

The chapter is divided into two main sections: section 5.1 presents the results attained using the carbon black as the fugitive material, whereas section 5.2 discusses the use of microcrystalline cellulose. Due to the different physical characteristics of the PZT-P505 powder batches (Table 3.2), which may influence the rheological behaviour, PZT-P505 batch 1 was used as the primary material when CB was used as the fugitive material. On the other hand, PZT-P505 batch 2 was employed when working with MCC.

5.1 Carbon black as the fugitive material

5.1.1 Rheological characterisations of the feedstocks: torque-rheometry

The feedstocks characterised and discussed in this subsection of the work are the feedstocks n$^{\circ}$ 2, 9-11 (PZT-P505 batch 1), and 14-16 listed in Table 3.4. The mixing and the rheological investigations were carried out at the same temperature of 120 $^{\circ}$C. This temperature was defined based on the results discussed for the thermoplastic binder material LDPE (Chapter 4).

As previously mentioned, a solid loading of 58 vol. % of PZT was selected for the primary feedstock. Since similar rheological behaviour among the different green body components is critical for shape retention during co-extrusion, to determine a carbon containing feedstock with similar flow characteristics as the primary feedstock, the unfilled LDPE composition was initially analysed (results from Chapter 4) and, subsequently, carbon black powder was progressively added (25, 35, and 45 vol. % CB containing feedstocks). The results of the rheological characterisation of the feedstocks by means of torque-rheometry are shown in Figure 5.1. The approach used to represent the results in this form was based on the following:

According to the Ostwald-de Waele power-law model **[She99]**, a rheological non-Newtonian model proposed for representing the polymeric flow behaviour, the relation between shear stress, τ , and shear rate, $\dot{\gamma}$, is given by equation (5.1):

$$\tau = \overline{K}\dot{\gamma}^n \tag{5.1}$$

where $\overline{K}$ reflects the consistency index of the material, and n is the power law index, giving a measure of the degree of non-Newtonian behaviour, with values smaller than unity showing shear-thinning characteristics. Assuming the torque-rheometer to be analogous to a concentric-cylinder rheometer (Figure 4.8), it was shown that there is a relation between the shear stress, the torque moment M of the mixer rotors and the dimensions k of the mixing chamber. Combining this relation with equation (5.1) yields:

$$M = \frac{\overline{K}\dot{\gamma}^n}{k} \tag{5.2}$$

Furthermore, taking into account the geometrical dimensions of the torque-rheometer (Figure 3.4), the apparent shear rate $\dot{\gamma}_{app}$ can be calculated from the angular velocity of the rotor, Ω, and the gap, y, between the rotor and the chamber wall. This is given by equation (5.3):

$$\dot{\gamma}_{app} = \frac{\Omega}{y} = \frac{2\pi rN}{60y} \qquad (5.3)$$

where r is the radius of the rotor and N is the speed of the rotors in revolutions per minute (rpm). Combining the equations (5.2) and (5.3) gives:

$$M = \frac{\overline{K}}{k}\left(\frac{2\pi rN}{60y}\right)^{n} \qquad (5.4)$$

By defining a rheological mixing constant $C(n)$:

$$C(n) = \frac{1}{k}\left(\frac{2\pi r}{60y}\right)^{n} \qquad (5.5)$$

equation (5.4) could be simplified as:

$$M = C(n)\overline{K}N^{n} \qquad (5.6)$$

Equation (5.6) enables one to plot the logarithm of torque (M) *versus* the logarithm of rotor speed (N) yielding a linear correlation, where the power law index n represents the slope of the curve. For representation of Figure 5.1, equation (5.6) was then combined with equation (4.4).

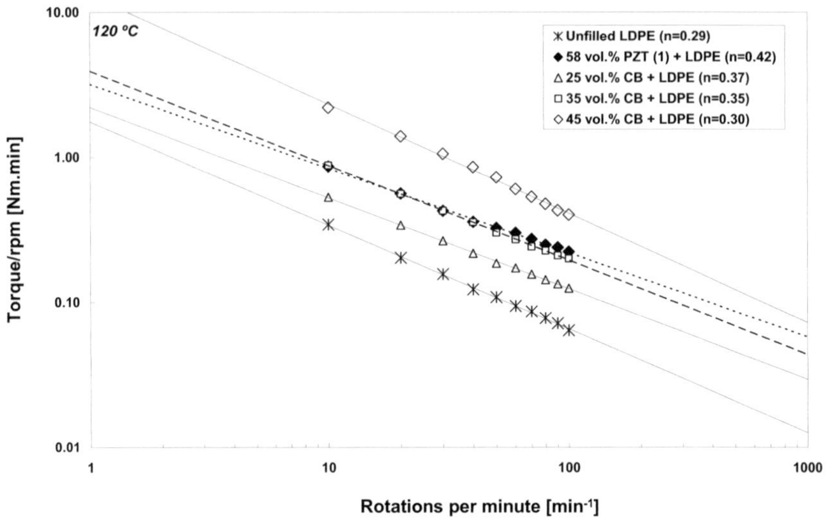

Figure 5.1: Torque/rpm as a function of rotations per minute for the different feedstocks characterised for the co-extrusion process. The power law index n is the negative slope of the curve.

The feedstock containing 35 vol. % of CB displayed similar rheological behaviour as the 58 vol. % PZT composition. All the feedstocks exhibited shear-thinning behaviour for the temperature and velocity range analysed, as the power law indices n were smaller than 1. The pseudoplasticity is a result of the alignment of the binder molecules at high shear rates. The lower degree of shear-thinning observed for the filled systems (PZT and CB containing feedstocks) in comparison to the unfilled feedstock (unfilled LDPE), is attributed to the fact that higher shear stresses are created with higher solid loading, since the inter-particle separation distance reduces. However, when comparing the CB containing feedstocks, the resistance to shear decreased when adding carbon black powder. This could be explained assuming a higher probability to ensue in a laminar flow in the presence of particles in filled polymer mixtures. When the particles are oriented by laminar flow, the resistance to shear decreases **[Wig01]**. Accordingly, the PZT feedstock should present a smaller power law index than the CB feedstocks, since it contains a greater amount of ceramic particles. Nevertheless, the size and shape of the particles are important factors to be considered when investigating the rheology of high solid loading mixtures.

5.1.2 Co-extrusion investigations

5.1.2.1 Monofilaments for solid-fibres production

Due to the fact that the 35 vol. % CB + LDPE feedstock indicates a similar torque behaviour at a given range of rotor speed as the 58 vol. % PZT (1) + LDPE feedstock (Figure 5.1), it was selected as the fugitive composition for further investigations. However, it is noted that the compositions have flow compatibility only for a given range of rotor speed. At lower rotor speeds (< 20 rpm), the CB feedstock exhibited a higher apparent viscosity (torque/rpm) than the PZT composition, while at higher rotor speeds (> 20 rpm) the opposite behaviour was observed. It is well known that for successful co-extrusion, viscosity matching is necessary between the processed materials. Any variation might result in a rearrangement of the interface and instabilities during flow. To explore these variations, different co-extrusion experiments were performed varying the velocity of the piston during extrusion, which provided different viscosity ratios between the materials. Table 5.1 summarizes the parameters used for the co-extrusion experiments.

Table 5.1: Parameters considered for the co-extrusion experiments for solid-fibres production, using the 58 vol. % PZT (1) + LDPE and 35 vol. % CB + LDPE feedstocks

Co-extrusion investigations	Capillary-rheometer		Torque-rheometer	
	Piston velocity [mm/min]	Apparent shear rate [1/s]	Correspondent range of rotors speed [1/min]	Apparent viscosity ratio range (PZT/CB)
Co-Ex 1	0.07	5	4 – 37	0.90 – 1.05
Co-Ex 2	0.26	20	15 – 149	0.98 – 1.16
Co-Ex 3	0.78	60	44 – 448	1.06 – 1.25
Co-Ex 4	1.30	100	73 – 747	1.10 – 1.30
Co-Ex 5	3.91	300	220 – 2241	1.19 – 1.40

Based on the data for the capillary-rheometer, given in Table 5.1, different speeds (piston velocities) were selected by means of equation (5.7) **[Mac94]**, where the velocity of the piston (V_{piston}) and the apparent shear rate at the wall of the die ($\dot{\gamma}_{app}$ [s^{-1}]) can be established.

$$\dot{\gamma}_{app} = \frac{4Q}{\pi R_d^3} \qquad (5.7)$$

with Q as the volumetric flow rate ($Q = V_{piston} \cdot A$ [mm^3/s], where V_{piston} is the velocity of the piston, A the area of the piston) and R_d the die radius [mm]. Taking into account the geometrical dimensions of the torque-rheometer (Figure 3.4), the correspondent range of rotors speed N was calculated using the aforementioned equation (5.3). The smallest gap between the rotor and the chamber wall provided the maximum rotor speed and the widest gap, the lowest one. The apparent viscosity ratio of the feedstocks is expressed as a torque/rpm ratio of the PZT compound (58 vol. % PZT (1) + LDPE) and the carbon black composition (35 vol. % CB + LDPE) for the corresponding apparent shear rate range. The co-extrusion experiments were carried out at the same temperature as the rheological characterisation (120 °C). Figure 5.2

shows the characterisation of the co-extruded monofilament composites. The dotted lines represent the theoretical diameter of the PZT inner fibres, i.e., exactly 24 times smaller than the preform ($\varnothing \sim 300$ µm). The cross section of the monofilament composites are imaged at a magnification of 200 x, which allowed the whole fracture surface to be seen.

(a)
Co-Ex 1 (V_piston = 0.07 mm/min)
Ext. V = 330 mm³
Ext. V = 562 mm³
Ext. V = 2144 mm³
Ext. V = 3295 mm³
Ext. V = 4037 mm³
Ext. V = 5586 mm³
PZT Fiber Area / Total Monofilament Composite Area
(b)
Co-Ex 2 (V_piston = 0.26 mm/min)
Ext. V = 120 mm³
Ext. V = 1041 mm³
Ext. V = 1431 mm³
Ext. V = 3847 mm³
Ext. V = 5037 mm³
Ext. V = 6322 mm³
PZT Fiber Area / Total Monofilament Composite Area
(c)
Co-Ex 3 (V_piston = 0.78 mm/min)
Ext. V = 509 mm³
Ext. V = 972 mm³
Ext. V = 2619 mm³
Ext. V = 3024 mm³
Ext. V = 3561 mm³
Ext. V = 5734 mm³
PZT Fiber Area / Total Monofilament Composite Area
Co-extruded volume [mm³]

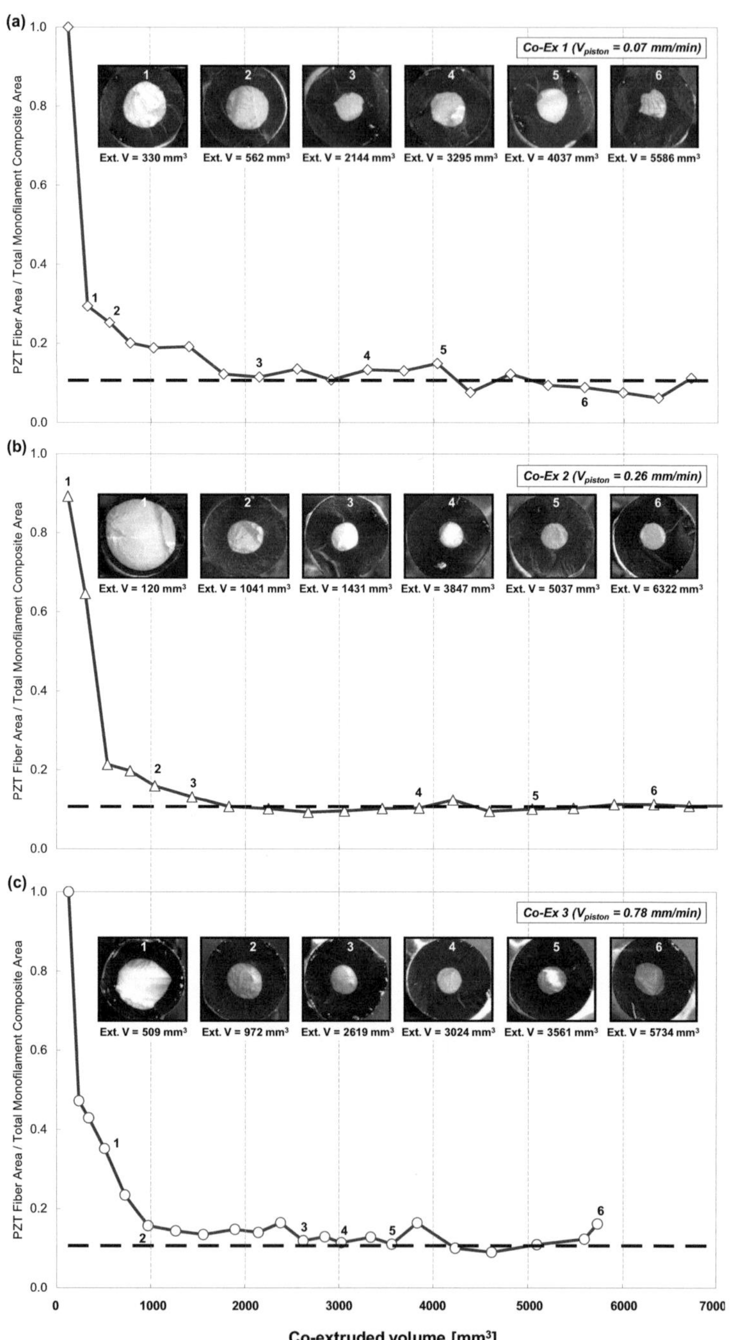

90

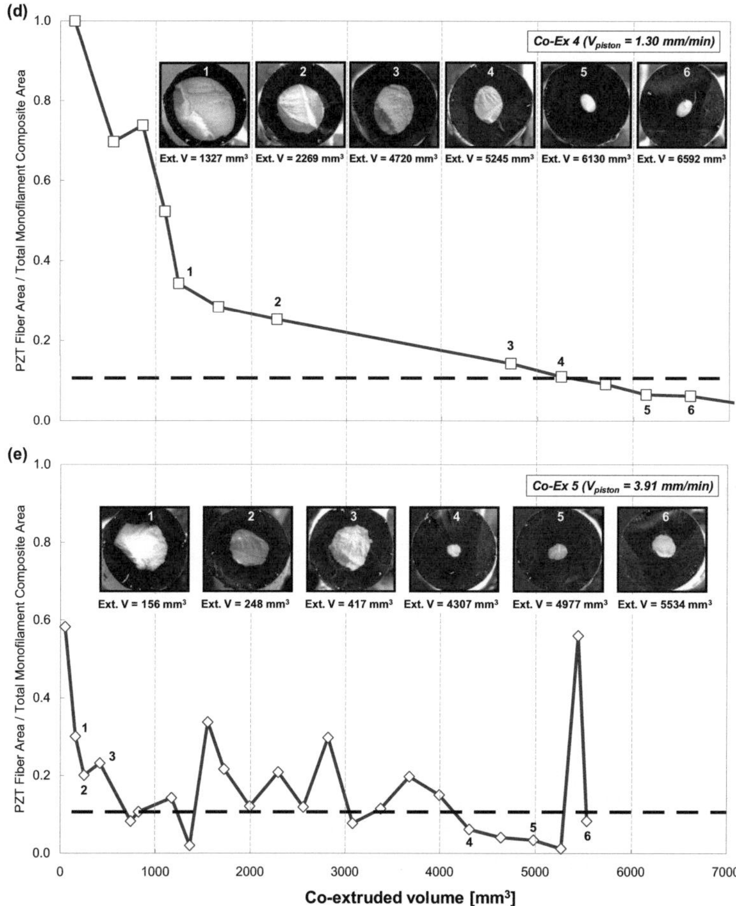

Figure 5.2: Co-extruded PZT fibre area and total monofilament composite area ratio as a function of the co-extruded volume for experiments carried out applying different piston velocities ((a) – (e)). The micrographs show the cross section (perpendicular to the extrusion direction) of the monofilament composites for different extruded volume (Ext. V).

At lower shear rates, i.e., Co-Ex 1, 2 and 3, the process starts to become stable only after a volume proportional to the volume of the cone part of the die was extruded (approximately 2400 mm^3). As the piston moves downward it is expected that the PZT, which lies in the centre of the preform composite, firstly comes out of the orifice of the die, because of the chosen geometry ratio. In addition, a static zone in which the extrudate is not squeezed out is produced at the die entry region **[Ben93]**. Hence, co-extruded monofilament composites with good qualities were observed only after the static region was formed and the flow became steady. At higher shear rates (Co-Ex 4 and 5) the stability was not achieved, since high extrudate velocities do not generate a steady flow. The results indicate that a successful co-extrusion with well shaped filament morphologies and defined interfaces between the co-extruded materials (Figure 5.3), can be obtained with the second co-extrusion experiment (Co-Ex 2, V_{piston} = 0.26 mm/min). With the applied range of viscosity ratios (Table 5.1), the co-extrusion process was stable after a short initialized period (co-extruded volume = volume of the cone part of the die) up to the end of the experiment, and a good correlation with the PZT fibre area and total monofilament composite area was obtained.

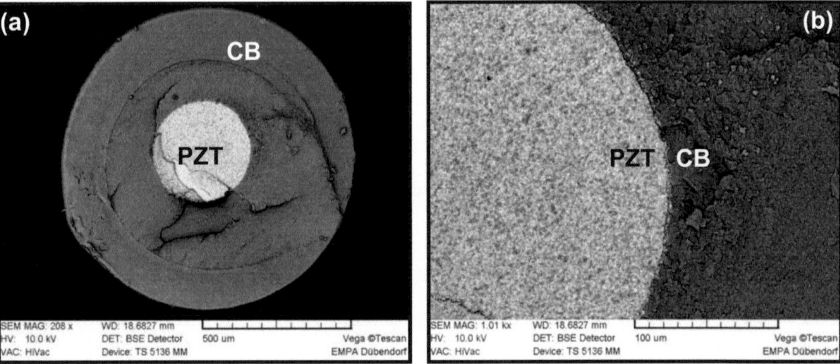

Figure 5.3: Micrographs of successfully co-extruded fibres (Co-Ex 2) indicating (a) well shaped fibre morphologies and (b) defined interfaces between the co-extruded materials.

For all other conditions, interface instabilities are observed, attributed to the viscosity mismatch between the primary and fugitive material. The Co-Ex 3, 4 and 5 were carried out with a carbon black/LDPE composition revealing a smaller viscosity than the primary material, as recommended by Schrenk et al. **[Sch78]**. The aim was to minimize the velocity difference between the co-extruded layers by using an inner layer with a higher viscosity. However, in this study, the CB material indicates only lower viscosities at high shear rates, which led to two drawbacks. First, experiments were performed applying a high piston speed. In accordance to Chen et al. **[Che08]**, co-extrusion of pastes should be carried out at a relatively low extrusion rate to achieve high quality products. Second, increasing the shear rate the difference in viscosity of the compositions was enhanced as well. This more pronounced difference in viscosity between the PZT and the CB feedstocks with increasing shear rate, can be attributed to the higher specific surface area of the carbon black powder used (Table 3.2). Feedstocks containing high specific surface area particles are more susceptible to surface phenomena, which can cause agglomeration between the particles. The agglomerates are porous structures formed due to the bonding of primary particles that absorbs a part of the polymer originally destined for the inter-particle separation, increasing the viscosity of the composition. By increasing the shear rate, there is a breakdown of the agglomerates, which releases the absorbed polymer that will contribute to the inter-particle separation, therefore decreasing the viscosity. Nevertheless, when comparing Co-Ex 1 with Co-Ex 3, where the difference in viscosity between the materials was less than 10% and the investigations were attained at lower shear rates, the results were quite similar to Schrenk et al. **[Sch78]**. Co-Ex 1 was carried out with the inner part having a lower viscosity than the CB material, while for Co-Ex 3 the opposite was observed and, as a result, interfacial flow instabilities were reduced, but not eliminated.

Furthermore, due to the fact that Co-Ex 3, 4 and 5 were performed when the carbon black/LDPE feedstock had a smaller viscosity than the primary material, it was reasonable to assume that the CB feedstock would be the first extrudate to come out of the die, encapsulating the inner PZT fibre. However, the die and preform geometry should be taken into account, as previously discussed. Moreover, for the case of the Co-Ex 3, the difference in viscosity between the materials was probably not enough for the encapsulation phenomena to occur. For the case of Co-Ex 4 and 5, due to the applied high speed, the residence time was not sufficient. However, the oval shape of the inner fibre obtained at the end of the Co-Ex 4 process is an indication of this instability. In summary, it is valuable to consider that in addition to the viscosity match between the materials, multilayer flow through a die is sensitive to extrusion rate.

5.1.2.2 Monofilaments for hollow-fibres production

Applying the same parameters of the successfully co-extruded feedstocks (Co-Ex 2, Table 5.1), monofilament composites for hollow-fibres production were processed. For that, the preform composite was the opposite than the one for the solid-fibres production (described in Chapter 3), but with the same formulation of the feedstocks (58 vol. % PZT (1) + LDPE and 35 vol. % CB + LDPE). Figure 5.4 shows the co-extruded CB fibre area (hollow) and total monofilament composite area ratio as a function of the co-extruded volume for the aforementioned co-extruded geometry, corroborating the successful co-extrusion. The monofilament composites are imaged at a magnification of 200 x, which allowed the whole fracture surface to be seen.

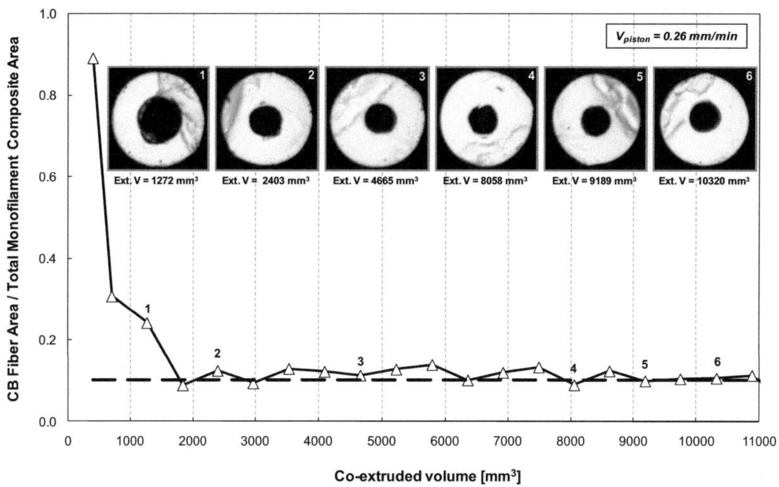

Figure 5.4: Co-extruded CB fibre area (hollow) and total monofilament composite area ratio as a function of the co-extruded volume for the co-extrusion experiments for hollow-fibres production. The micrographs show the cross section (perpendicular to the extrusion direction) of the monofilament composites for different extruded volume (Ext. V).

5.1.3 Rheological characterisations of the feedstocks: viscosity models

For the application of the different viscosity models (Eq. (2.9) – (2.12)) it was considered that the relative viscosity η_r is the quotient of the torque/rpm of the feedstock (η_s) to the torque/rpm of the binder system (η_o). The maximum solids content ϕ_{max} was calculated by plotting the reciprocal torque as a function of the ratio between the binder and the filling content (PZT or CB powder) (Figure 5.5), for the experiments carried out at 20 rpm and 120 °C. By extrapolating the data to zero reciprocal torque (i.e. an infinite magnitude of torque), the maximum powder loading could be determined **[Cle07b]**.

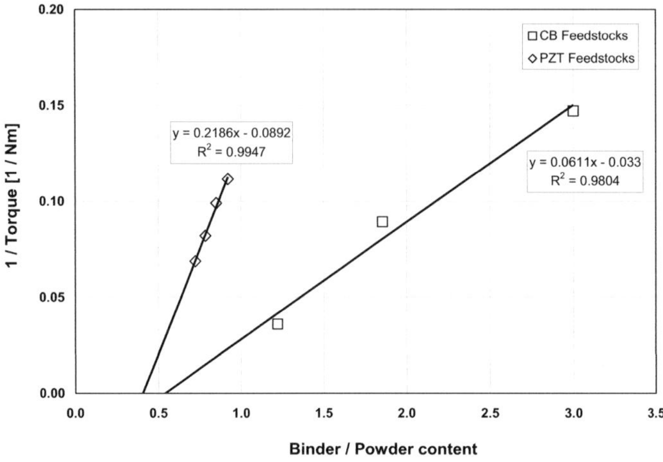

Figure 5.5: Reciprocal torque as a function of the ratio between the binder and the filling content (PZT-P505 batch 1 or CB powder) at 20 rpm and 120 °C.

A maximum powder loading (ϕ_{max}) of 71 vol. % was determined for the PZT feedstock, whereas for the CB system this value was lowered to 64 vol. %. The parameter ϕ_{max} depends on the arrangement of the particles, which in turn is determined by the particle shape, particle size distribution and shear flow **[She99]**. In the densest lattice packing, a packing fraction of 0.74 can be achieved when considering ideal monodisperse spherical particles. Donev et al. **[Don04]** demonstrated that ellipsoid particles, depending on their aspect ratio, can be randomly packed up to 0.71. Polydispersity yields a higher value of ϕ_{max} because the smaller particles can fit into the gaps between the bigger ones. The shear-flow dependence is due to the fact that the maximum attainable concentration is affected by the degree of agglomeration of the suspended particles. Agglomerates are porous structures formed due to the bonding of primary particles, which may absorb part of the material originally destined to fill the voids between particles. In this study, the ϕ_{max} difference observed between the two feedstocks is attributed to the binder system of the compositions and to the high specific surface area of the CB particles. Besides the high molecular-weight binder (LDPE), the PZT feedstock is composed of a low molecular-weight surfactant (stearic acid), which creates a better compatibility between the polymer and the inorganic particles. Additionally, it hinders the re-agglomeration of the particles after they have been broken apart by mechanical shearing. With regard to the specific surface area, compositions containing high SSA particles are more susceptible to surface phenomena, which can result in agglomeration between particles, thereby reducing ϕ_{max}.

Figure 5.6 and Figure 5.7 indicate the relative viscosity as a function of the solids content for PZT and CB feedstocks, respectively. Considering the ϕ_{max} calculated from Figure 5.5, plots of Chong, Frankel–Acrivos, Quemada and Krieger–Dougherty equations and the correspondent fit stability index (R^2) are additionally represented.

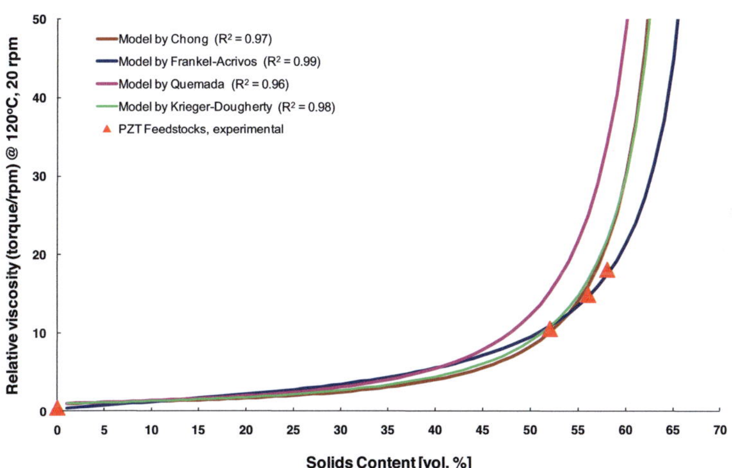

Figure 5.6: Relative viscosity (η_r) as a function of solids content (ϕ) for PZT feedstocks (PZT-P505 batch 1) at 20 rpm and 120 °C compared with plots of Chong, Frankel-Acrivos, Quemada and Krieger-Dougherty equations considering $\phi_{max} = 0.71$, $C' = 1.125$ and $\eta = 2.5$.

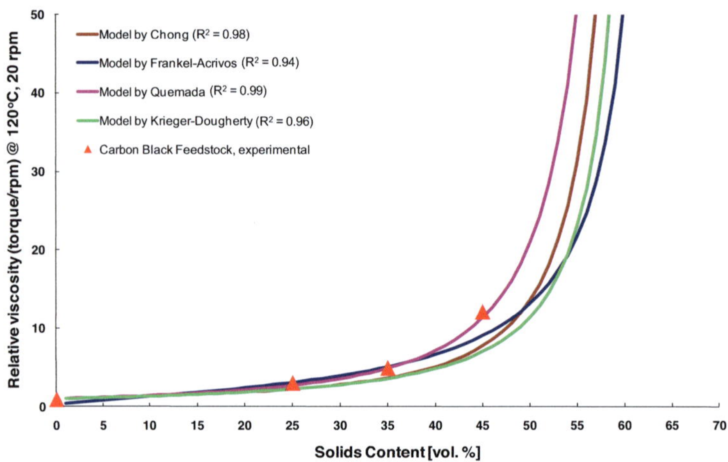

Figure 5.7: Relative viscosity (η_r) as a function of solids content (ϕ) for CB feedstocks at 20 rpm and 120 °C compared with plots of Chong, Frankel-Acrivos, Quemada and Krieger-Dougherty equations considering ϕ_{max} = 0.64, C' = 1.125 and η = 2.5.

For the PZT feedstocks (Figure 5.6), the model described by Frankel-Acrivos, when using the calculated ϕ_{max} = 0.71, showed the best agreement with the experimental data, depicted from the R^2 value close to 1. Considering the data points, a better fitting of the Chong model would be possible estimating a ϕ_{max} = 0.72. For the Quemada equation, a better prediction of viscosity is achieved considering ϕ_{max} = 0.76, while for the Krieger-Dougherty model a ϕ_{max} = 0.73 or an intrinsic viscosity equal to 2.6 should be taken into account. Hence, due to the value of ϕ_{max} estimated by the Quemada equation, this model would be not feasible for the PZT feedstocks studied here, whereas the Chong and Krieger-Dougherty models would still be suitable.

For the CB feedstocks (Figure 5.7), the model predicted by Quemada (ϕ_{max} = 0.64, R^2 = 0.99) better represents the experimental data. Using a critical volume fraction of 0.56 for the Chong equation, a superior fit quality would be attained. For a suitable adjustment of the Krieger-Dougherty model, the value for the ϕ_{max} should be 0.47 or an intrinsic viscosity equal to 3.3. For a superior fitting of the Frankel-Acrivos model, a critical volume fraction of 0.61 should be applied. Alternatively, a superior fitting would be achieved using the calculated $\phi_{max} = 0.64$ in combination with a $C' = 1.5$, which would be a suitable value, since the increase in C' suggests the deviation from spherical particles.

It is noted that for an increase of the fit quality of the Krieger-Dougherty model both feedstocks should have a higher intrinsic viscosity than Einstein's calculated one ($\eta = 2.5$). The intrinsic viscosity of a suspension is a function of the particle geometry. A deviation from spherical particles leads to a higher intrinsic viscosity due to the increasing disturbance of the flow field. In addition, the value of η increases with the degree of agglomeration **[Oka00]**. Accordingly, the superior η expected for CB containing feedstocks ($\eta = 3.3$), when compared with PZT compositions ($\eta = 2.6$), is explained due to the higher SSA of the CB particles.

For the case of PZT feedstocks, the semi-empirical viscosity models studied is useful for understanding the solids loading-viscosity relation, avoiding time-consuming and cost-intensive investigations into the maximum accessible load. To achieve low shrinkage during debinding and sintering and high-quality products, it is necessary for the PZT feedstock formulation to be as close to the critical volume fraction of the solids as possible. However, it can be seen from these models that the relative viscosity rises rapidly as this value is approached, which decreases the processing ease. This sharp rise occurs because the viscosity of highly loaded systems is dominated by the particle-particle interactions.

Considering CB containing feedstocks, these models are functional in estimating the amount of CB in the fugitive composition presenting similar rheological behaviour to the primary compound, at a given shear rate and temperature, a prerequisite for successful co-extrusion. This is shown in

Figure 5.8, where the torque/rpm of the CB feedstock (η_s) as a function of solids content (ϕ) is plotted. Additionally, the value of torque/rpm for the 58 vol. % PZT (1) + LDPE feedstock at 120 °C and 20 rpm is represented. As the Quemada equation has shown the best agreement for the CB data points, this model is the one represented. It is observed that the Quemada model predicts a solid content of 35 vol. % of CB in the fugitive feedstock in order to achieve a similar viscosity to the primary feedstock (58 vol. % PZT), which is in good agreement with the results discussed so far.

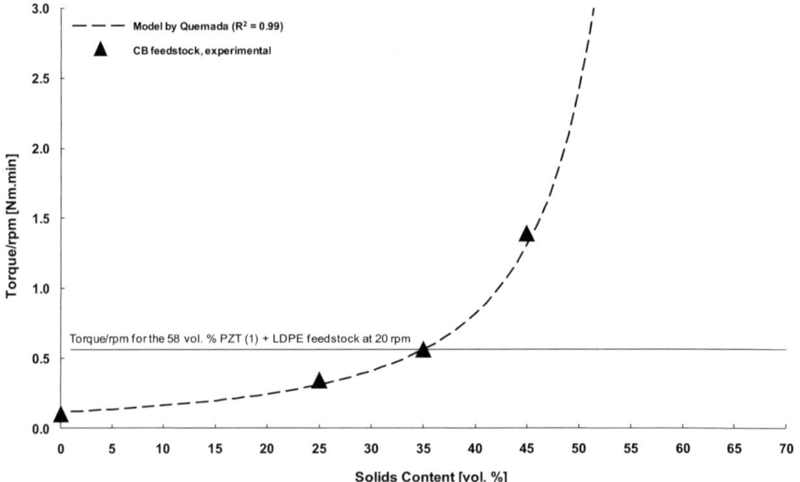

Figure 5.8: Torque/rpm (η_s) as a function of the solids content (ϕ) for CB feedstocks at 20 rpm and 120 °C compared with plot of Quemada equation considering ϕ_{max} = 0.64. The value of torque/rpm for the 58 vol. % PZT (1) + LDPE feedstock is represented.

5.1.4 Binder and carbon black removal

The binder and carbon black removal (debinding step) of the co-processed materials (monofilaments for solid and hollow-fibres production) were performed by thermal degradation. Based on TG-analysis of the carbon black feedstocks (Table 5.2), the debinding step was carried out in oxygen atmosphere in order to accelerate the decomposition of the organic parts and to minimize effects on densification of the PZT-ceramic. However, from Table 5.2 it can be seen that even in pure O_2 atmosphere, 2% of residue is still left at 600 °C, the temperature when PZT starts sintering. In view of that, it is reasonable to assume that the elevated temperature decomposition of the fugitive feedstock may negatively affect the densification process of the inner fibre as well as it may degrade the physical properties of the PZT-ceramic. Debinding in air would require higher temperatures compared to pure O_2 and results in a significant higher residue (23%) at 600 °C.

Table 5.2: Information derived from TGA measurements (1 °C/min) for the 35 vol. % CB + LDPE feedstock

Condition	T_{onset} [°C]	T_{max} [°C]	Residue at 600 °C
in air atm.	275	615	23 %
in O_2 atm.	225	580	2 %

Moreover, it is worth mentioning that after the debinding step of the co-processed monofilament composites for solid-fibres production, it was verified that the inner fibres were fractured in pieces of about 8 mm. However, this was not the case for monofilament composites for hollow-fibres production. Again, this is a drawback probably originated due to the high temperature decomposition of the outer layer (carbon black feedstock). Figure 5.9 schematically illustrates the cross-section (longitudinal direction) of the monofilament composites for the solid- and hollow-fibres production.

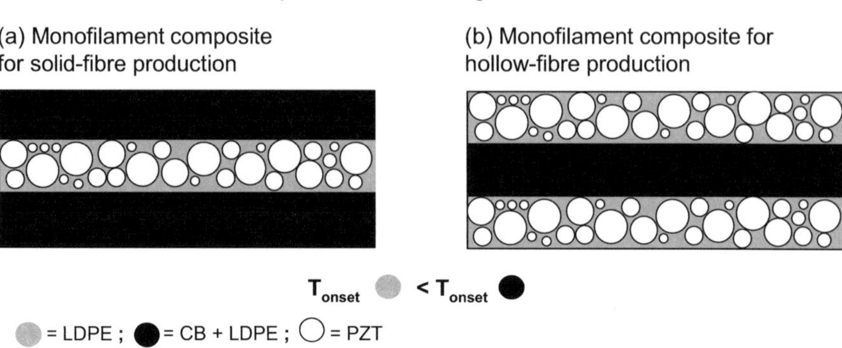

Figure 5.9: Schematic diagram illustrating the cross-section (longitudinal direction) of the monofilament composites for the (a) solid and (b) hollow-fibres production. The temperature at which weight loss onset occurs (T_{onset}) is lower for the primary feedstock (PZT + LDPE) than for the fugitive feedstock (CB + LDPE).

For the case of monofilament composites for solid-fibres production (Figure 5.9 (a)), as heat is applied, the volatile products originated by the degradation of the low density polyethylene binder in the inner PZT fibre diffuse to the green body surface (PZT fibre) in order to be eliminated. However, the outer layer (CB feedstock) prevents out-gassing and internal stresses will be generated in the inner fibre. If these stresses exceed some critical value, fracture of the inner fibre occurs. On the other hand, for the case of monofilament composites for hollow-fibres production (Figure 5.9 (b)), as heat is applied, the organic materials of the outer layer (PZT feedstock) can be eliminated, creating paths and thus facilitating the removal of the volatile substances of the inner layer.

5.2 Microcrystalline cellulose as the fugitive material

The processing steps for the solid and hollow-fibres production using microcrystalline cellulose as the fugitive filler are based on the results of the successful co-extrusion with carbon black as the fugitive substance. The feedstocks characterised and discussed in this subsection are the feedstocks n° 12 (PZT-P505 batch 2), and 17-21 listed in Table 3.4.

5.2.1 Rheological characterisations of the feedstocks: torque-rheometry and viscosity models

A primary condition for a successful co-extrusion is the adjustment of a similar rheological behaviour of the fugitive feedstock and the primary feedstock for a given shear rate and temperature, as discussed in the section 5.1. In order to identify the MCC feedstock with a similar apparent viscosity as the 58 vol. % PZT (2) + LDPE feedstock, compositions containing 20, 30, 40 and 50 vol. % of MCC (feedstocks n° 17-21, Table 3.4) were rheologically characterised applying torque-rheometry. The experimental data were compared to the models of Chong, Frankel-Acrivos, Quemada and Krieger-Dougherty. The maximum solids content ϕ_{max} was calculated by plotting the reciprocal torque as a function of the ratio between the binder (LDPE + St. Acid) and the filling (MCC) content (Figure 5.10), as discussed for CB (Figure 5.5). The results presented are the data collected at 20 rpm and 120 °C, since at these conditions successful co-extrusion of CB and PZT feedstocks was attained.

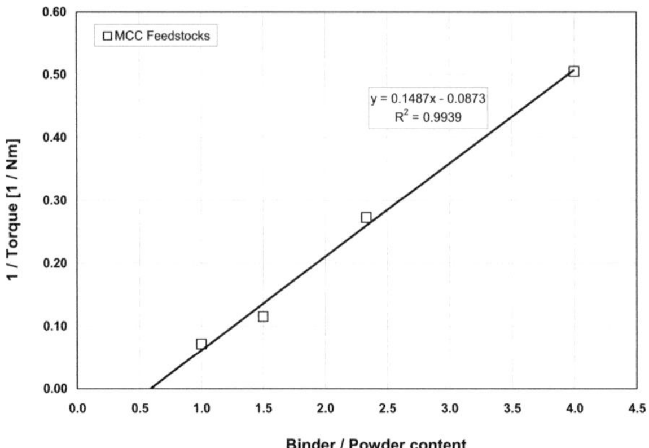

Figure 5.10: Reciprocal torque as a function of the ratio between the binder (LDPE + St. Acid) and the powder content (MCC) at 20 rpm.

From Figure 5.10 a maximum powder loading (ϕ_{max}) of 63 vol. % has been calculated. Applying this value to the aforementioned viscosity models, it was verified that the Frankel-Acrivos equation showed the most satisfactory description to the solids-loading-viscosity data for the MCC feedstocks (R^2 = 0.94). In view of that, this model was used to calculate the solid loading of the fugitive feedstock (Figure 5.11).

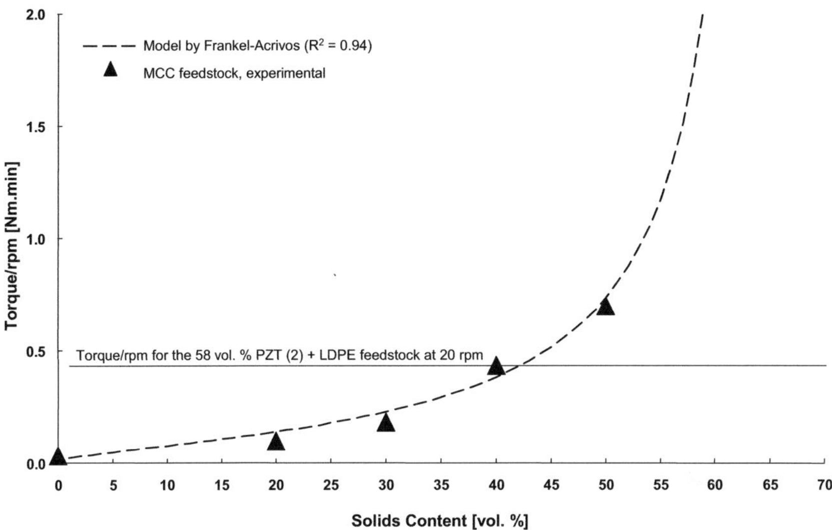

Figure 5.11: Torque/rpm (η_s) as a function of solids content (ϕ) for MCC feedstocks at 20 rpm and 120 °C compared with plot of Frankel-Acrivos equation considering ϕ_{max} = 0.63 and C' = 1.125. The value of torque/rpm for the 58 vol. % PZT (2) + LDPE feedstock is represented.

Solving the Frankel-Acrivos equation (Figure 5.11), the feedstock containing 41 vol. % of MCC should display similar torque/rpm as the 58 vol. % PZT (2) + LDPE feedstock, at 20 rpm and 120 °C. In order to confirm the prediction of the model, the 41 vol. % of MCC + LDPE feedstock was prepared and rheologically characterised. Figure 5.12 indicates the torque/rpm as a function of rotations per minute for the 58 vol. % PZT (2) + LDPE and 41 vol. % MCC + LDPE feedstocks, pointing out the viscosity matching between those feedstocks at 20 rpm. Therefore, these feedstocks were selected for the co-extrusion of the PZT-based fibres using MCC as the fugitive filler.

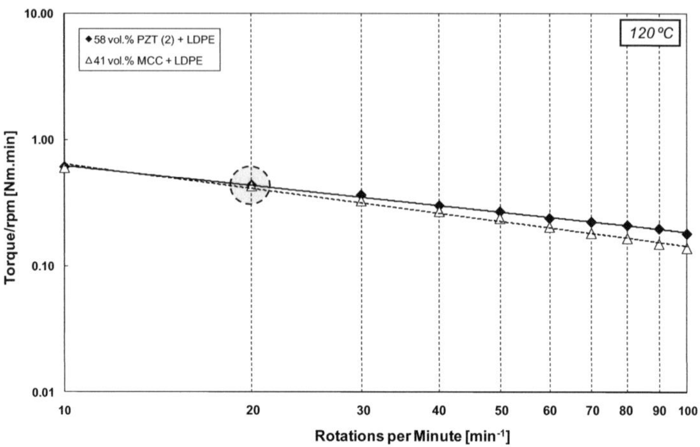

Figure 5.12: Torque/rpm as a function of rotations per minute for 58 vol. % PZT (2) + LDPE and 41 vol. % MCC + LDPE feedstocks.

5.2.2 Solid and hollow-fibres production

Based on the successful co-extrusion of preform composites formed by PZT and CB feedstocks for solid and hollow-fibres production, the same parameters (Co-Ex 2, Table 5.1) were applied for the fibres production using the MCC based feedstock (41 vol. % MCC + LDPE) as the fugitive part.

5.2.3 Binder and microcrystalline cellulose removal

The co-extruded monofilament composites using MCC as the fugitive filler were decomposed in air (Table 3.5). It is important to mention that no residue could be detected on the fibre surface after debinding up to 600 °C. However, after debinding of the monofilament composites for solid-fibres production, the fracture of the fibres still occurred (in pieces of about 8 mm).

Figure 5.13 indicates the decomposition behaviour (derivate thermogravimetry) of the PZT and MCC feedstocks used for the co-extrusion process. Additionally, the MCC powder decomposition profile is shown. In order to be compared with the MCC feedstock, the mass of PZT ceramic in the PZT feedstock has been normalized out.

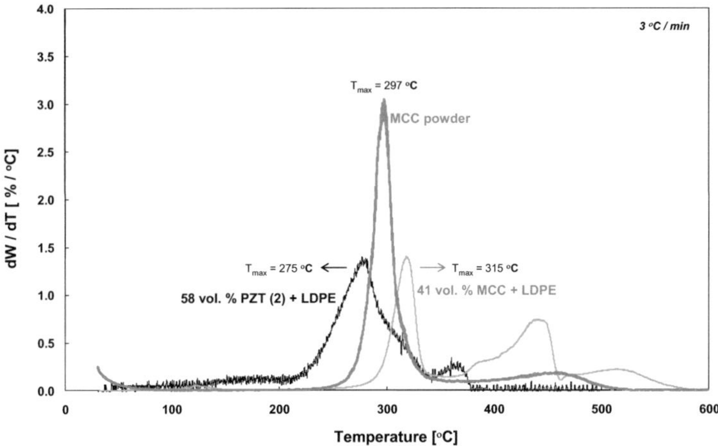

Figure 5.13: Decomposition profile (in air) of the pure MCC powder and of the PZT and MCC feedstocks used for the co-extrusion process. *Note: The PZT feedstock has been normalized to compare with the MCC feedstock.*

Figure 5.13 shows that the decomposition of the MCC feedstock is retarded when compared to the pure MCC powder. Additionally, it can be seen that the PZT feedstock starts decomposing earlier than the MCC feedstock (outer layer for the case of solid-fibres production). Therefore, it is reasonable to assume that, as mentioned for the CB processed fibres, the fracture occurred due to the higher onset temperature decomposition of the outer layer when compared to the T_{onset} of the inner layer. In order to optimize the burnout process, and subsequently avoid deterioration of the final properties of the ceramic material, a proper choice of the composition of the outer layer feedstock should be explored in a future work. It is well known that low molecular weight vehicles, such as waxes and polyethylene glycols, are added in the composition of injection moulded parts **[Ree95]**. These materials are removed earlier by, for example, solvent extraction, opening up penetrating pore channels, which significantly facilitates the subsequent removal of the higher molecular weight binder. However, for the case of co-extrusion, in which the rheological behaviour of the feedstocks plays a major role, the use of such materials would alter their rheological performance. Thus, for the fugitive feedstocks studied here, the use of a higher molecular weight LDPE (lower melt flow index) than the utilized one (Table 3.3) in combination with low molecular weight binders would be an alternative to optimize the binder burnout and, at the same time, reveal the same rheological behaviour as the primary feedstock (58 vol. % PZT).

5.3 Summary

The objective of the work in this chapter was to process solid and hollow PZT-based fibres by using the co-extrusion process. The feedstocks were prepared based on the materials selected in Chapter 4: the piezoelectric PZT-P505 as the primary material, carbon black (first subsection) or microcrystalline cellulose (second subsection) as the fugitive substance and low density polyethylene as the thermoplastic binder.

Assuming the torque-rheometer to be analogous to a concentric-cylinder rheometer, the rheological behaviours of a PZT/LDPE feedstock (primary feedstock) and CB/LDPE mixtures (fugitive feedstock) were analysed. Additionally, the rheological properties of these feedstocks were discussed in view of the rheological models derived by Chong, Frankel-Acrivos, Quemada and Krieger-Dougherty. These viscosity models appeared to be useful in estimating the solid loading of the fugitive filler at a given viscosity. Based on the rheology of the compounds, a feedstock containing 58 vol. % of PZT and another containing 35 vol. % of CB were selected for the co-extrusion of monofilament composites for solid-fibres production. However, the difference in pseudoplasticity of these compositions (PZT: n = 0.42, CB: n = 0.35) indicated flow compatibility between the feedstocks only for a narrow range of shear rates. Therefore, after assembling the preform composites, 5 co-extrusion experiments were carried out by deliberately employing different viscosities between the materials. In view of that, structure instabilities could be detected on green monofilament composites. Successful co-extrusion (working with R = 24:1) with well preserved fibre morphologies and defined interfaces between the co-extruded materials, was obtained for a viscosity ratio between the feedstocks ranging from 0.98 to 1.16. Outside of this range, interface distortions were observed. Despite the viscosity mismatch between the compounds, interfacial flow instabilities were reduced when the lower viscosity material was in contact with the die wall, considering the viscosity difference between the materials not higher than 10%. In general, co-extrusions carried out at lower shear rates were more satisfactory when compared to high piston speeds, due to the fact that high velocity extrusion may generate unsteady flow conditions. Furthermore, it could be demonstrated that the torque-rheometer is a practical tool to identify the rheology requirements for a successful co-extrusion of ceramic-polymer compounds.

Based on the successful co-extrusion of monofilament composites for solid-fibres production, monofilament composites to obtain hollow-fibres were processed. The binder and fugitive material removal of both processed filament geometries (solid and hollow) were carried out by thermal decomposition in an O_2 atmosphere. However, deleterious residual carbon was verified at the surface of the solid-fibres, attributed to the high decomposition temperature (above 600 $^{\circ}$C) of the carbon black feedstock. In addition, the solid-fibres appeared fractured after the binder removal process. This fact was attributed to the higher temperature decomposition of the outer layer, when compared to the organic material present in the inner part.

The second part of this chapter discussed the results obtained using microcrystalline cellulose as the fugitive filler in the fugitive feedstock. The processing steps for the solid and hollow-fibres production were founded on the successful co-extrusion of PZT and CB-feedstocks. Using the viscosity model described by Frankel-Acrivos (ϕ_{max} = 0.63), a MCC feedstock with corresponding viscosity at a given shear rate as the primary feedstock (58 vol. % PZT) could be determined (41 vol. % MCC). Subsequent to the co-extrusion and binder and fugitive material removal steps, solid and hollow-fibres were successfully obtained without residue at the surface of the solid -fibres. However, the fracture of the solid -fibres during the stage of binder/fugitive removal persisted. The addition of low molecular weight vehicles in the composition of the outer layer was suggested in order to facilitate the removal of the organic material presented in the inner layer.

5.4 Outlook

In view of the successful developed co-extrusion process of monofilament composites for solid and hollow-fibres production, an additional complex preform geometry, named half-moon shape, was challenged to be co-extruded. Figure 5.14 compares the cross-section geometries (radial direction) of the preforms for solid and hollow-fibres production (Figure 5.14 (a)) with the one for half-moon-filaments production (Figure 5.14 (b)).

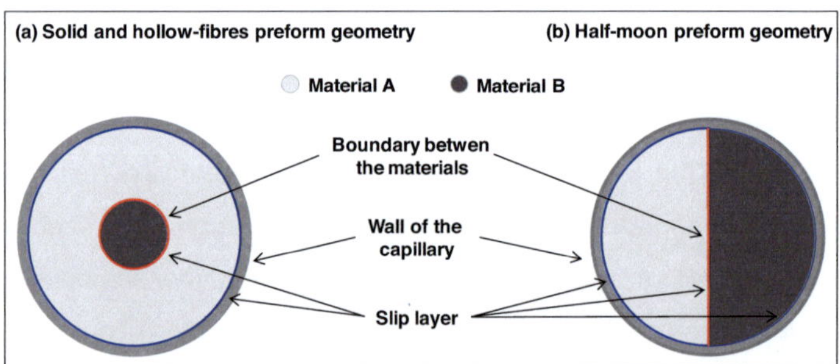

Figure 5.14: Schematic diagram illustrating the cross-section geometries (radial direction) of the preforms for the (a) solid and hollow-fibres production and (b) half-moon-filament production.

When evaluating the different preform geometries, it is observed that, for both cases, there is a boundary between the materials A and B, where, in case of different wall slip velocities of the materials, a relative velocity between the materials during extrusion may be generated. On the other hand, while only one material is in contact with the wall of the capillary for the solid and hollow-fibres geometry (Figure 5.14 (a)), this is verified for both materials when working with the half-moon geometry (Figure 5.14 (b)). For the solid and hollow-fibres production, it was demonstrated that successful co-extrusion is feasible when the apparent viscosity of the different materials to be co-extruded is matched and, additionally, when a low piston velocity during extrusion is applied. Nevertheless, it is reasonable to assume that, besides the viscosity matching between the materials, for the half-moon shaped preform the wall slip effect plays an important role during the co-extrusion process. Based on this assumption, for the successful co-extrusion of the half-moon preform geometry, a proper rheological characterisation of the materials to be co-extruded should involve the determination not only of the viscosity as a function of the shear rate but additionally of the slip velocity as a function of proper dependent variables.

6 Results and discussion - co-extrusion of PZT fibres: microstructure and electromechanical properties

The current chapter discusses the final properties of the sintered PZT solid-fibres. In order to establish the developed co-extrusion process, the microstructure and the electromechanical response of the co-extruded PZT solid-fibres are compared with the PZT solid-fibres with similar diameter (~ 250 µm after sintering) produced by the conventional extrusion technique (Chapter 4). The thermoplastic binder used for both processing routes was the low density polyethylene. With regard to the co-extruded fibres, the characterisation of the fibres attained using both carbon black (CB co-extruded) and microcrystalline cellulose (MCC co-extruded) as the fugitive fillers are presented. The final properties of the co-extruded hollow-fibres were not characterised within this work. Nevertheless, for illustration purposes, Figure 6.1 shows the fracture surfaces of successfully sintered hollow-fibres.

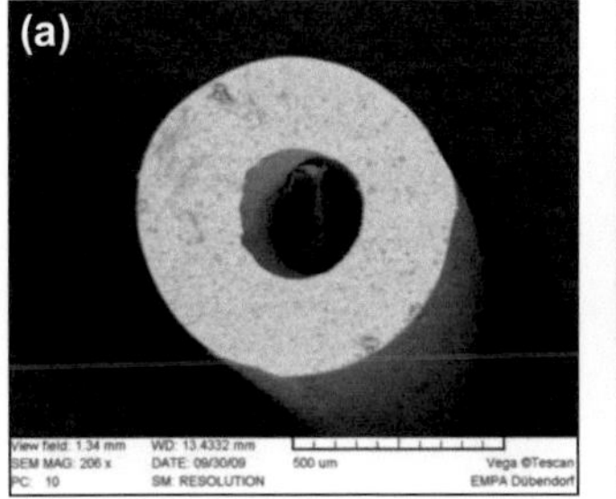
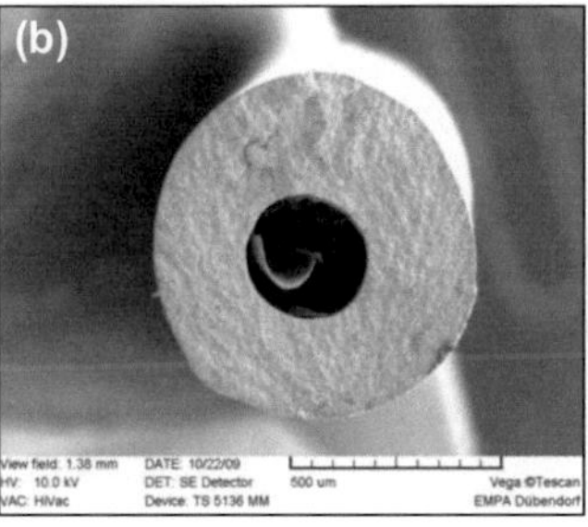

Figure 6.1: Fracture surfaces of sintered hollow PZT fibres obtained by the developed thermoplastic co-extrusion process: (a) using MCC as the fugitive material and (b) using CB as the fugitive material.

Figure 6.2 presents the fracture surfaces of sintered PZT solid-fibres obtained by both extrusion and co-extrusion techniques. Their microstructural parameters are listed in Table 6.1. For the extruded as well as for the MCC co-extruded fibres, the porosity was measured to be ~ 2%, i.e., a density of 98% of the theoretical density was achieved, revealing an almost complete densification process. A slightly higher porosity was observed for the CB co-extruded fibres (~ 5%), a value still suitable for piezoelectric fibre materials **[Hei09]**. The grain size did not display a significant dependence on the processing route. A core-shell microstructure (changes across the fibre radius) was not observed for any of the fibres, although a wide grain size distribution was found - grains between 3 and 11 µm were found for both processing routes. Figure 6.3 illustrates micrographs of etched extruded and co-extruded sintered PZT fibres. The slightly higher porosity of the CB co-extruded fibres can be evidenced by these images.

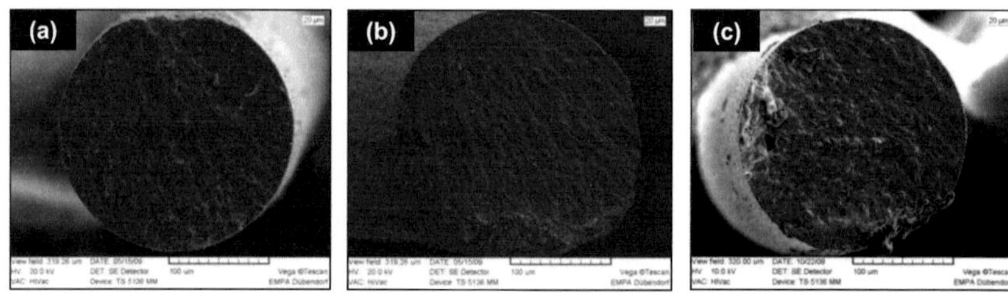

Figure 6.2: Fracture surfaces of sintered (a) extruded, (b) CB co-extruded and (c) MCC co-extruded PZT fibres.

Table 6.1: Microstructure properties of the extruded and co-extruded fibres

Fibre	S_{theo} [%]	Porosity [%]	Grain size [µm]	S_{calc} [%]
Extruded		1.94 ± 0.48	7.9 ± 1.8	16.06 ± 0.02
CB co-extruded	16.60	5.25 ± 1.65	7.3 ± 1.9	15.09 ± 0.06
MCC co-extruded		2.09 ± 0.37	7.5 ± 1.9	16.02 ± 0.01

Note: S_{theo} and S_{calc} refer to the theoretical and calculated shrinkage, respectively, as defined in Chapter 3.

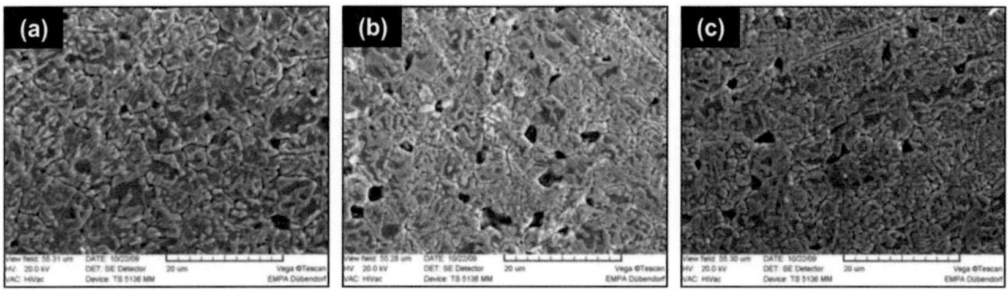

Figure 6.3: Etched images of sintered (a) extruded, (b) CB co-extruded and (c) MCC co-extruded PZT fibres.

Figure 6.4 shows the X-ray diffraction pattern for the different processed fibres and for the different measurement setups (fibre bulk material and fibre surface) in the range of $43° \leq 2\Theta \leq 46°$; Figure 6.5 illustrates the characteristic reflex group of a morphotropic PZT-based material **[Kun05]**. In this study the (200) reflex group was selected for examination due to its higher intensity when compared to other peaks **[Fer95]**. In addition, due to the small differences between the tetragonal (T) and rhombohedral (R) cell parameters, many of the diffraction lines in the coexistence region overlap to some degree. This is simplified by analysing a (h00) plane since the lines of the cubic system split only into two lines in the T phase and do not split at all for the R structure. Therefore, for compositions close to the desired nearly temperature-independent morphotropic phase boundary (MPB), where both T and R phases coexist, three distinct peaks are produced, i.e. $(002)_T$, $(200)_R$ and $(200)_T$ **[Fer95]**.

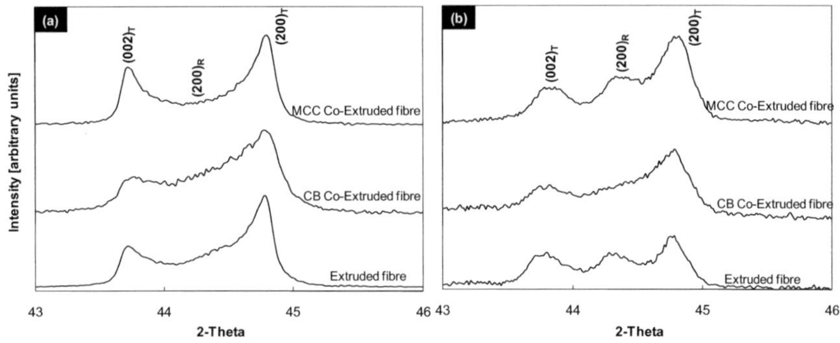

Figure 6.4: X-ray diffraction pattern of extruded and co-extruded PZT fibres (a) bulk materials and (b) fibre surfaces (T = tetragonal phase; R = rhombohedral phase). *Note: Extruded and CB co-extruded fibres were produced using PZT-P505 batch 1, while MCC co-extruded fibres were produced using PZT-P505 batch 2.*

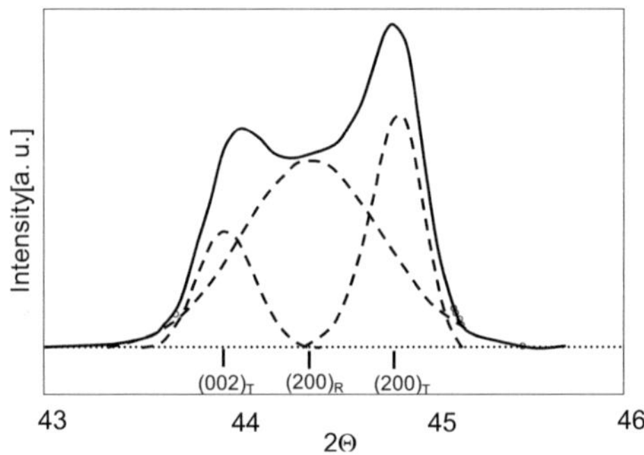

Figure 6.5: Schematic reflexes groups for morphotropic PZT **[Kun05]**.

For all processed fibres the coexistence of the two ferroelectric phases can be observed with the characteristic (002) and (200) doublet of the tetragonal phase and the (200) single peak of the rhombohedral phase (Figure 6.4). The X-ray patterns indicate a predominant tetragonal phase and a lower amount of rhombohedral phase for bulk material of extruded and MCC co-extruded fibres. In contrast, CB co-extruded fibre reveals patterns typical for PZT materials with equal quantitative of both phases. Analysis of the fibre surface (Figure 6.4 b) shows that the pattern of the CB co-extruded fibre remains nearly unchanged compared to the bulk, whereas the rhombohedral peak became more pronounced for MCC co-extruded and extruded fibres indicating a higher rhombohedral phase content at the surface.

It is worth noting that the CB and MCC co-extruded fibres were produced using different PZT powder batches, as previously mentioned in Chapter 5. Nevertheless, this would not explain the difference in phase composition attained for those different co-extruded materials, as a difference in phase composition is also verified between the extruded and the CB co-extruded fibres, and both fibres were produced using the same PZT powder batch. The co-extruded and extruded fibres were sintered inside the same closed alumina crucibles (Chapter 3), neglecting the feasibility that the variations in phase composition are related to sintering conditions.

One reasonable explanation due to changes in phase compositions (bulk to fibre surface) observed for MCC co-extruded and extruded fibres, is the assumption of a gradient in phase fraction across the fibre radius. This statement is in good agreement with the work of Heiber et al. **[Hei07]**, where it was shown by microprobe examinations that the chemical (Zr and Ti ions) and thus the phase composition vary across the radius of thin extruded PZT fibres sintered in PbO-enriched atmosphere. They reported that at the fibre surface of 250 μm diameter extruded PZT fibres, the Zr^{4+}/Ti^{4+} ratio was shifted to slightly higher Zr^{4+} concentrations, which in turn was characterised by a predominant rhombohedral phase. This was explained assuming that the PbO liquid phase, formed during sintering (**[Kin83a]**, **[Kin83b]**, **[Ham98]**), contains titanium oxide. Due to the higher partial pressure in the liquid state, it was then assumed that Ti^{4+} diffuses out of the fibres and evaporates at the fibre surface. However, this was not investigated within this work.

Figure 6.6 shows the electromechanical performance of the extruded and co-extruded fibres. Strain data is presented in terms of relative strain, given in percentage. On the basis of the recorded butterfly curves, the maximum strain (S_{max}), the remanent strain (S_{rem}), the coercive field (E_c) and the effective high field piezoelectric coefficient (d_{33}^*) were determined (Table 6.2).

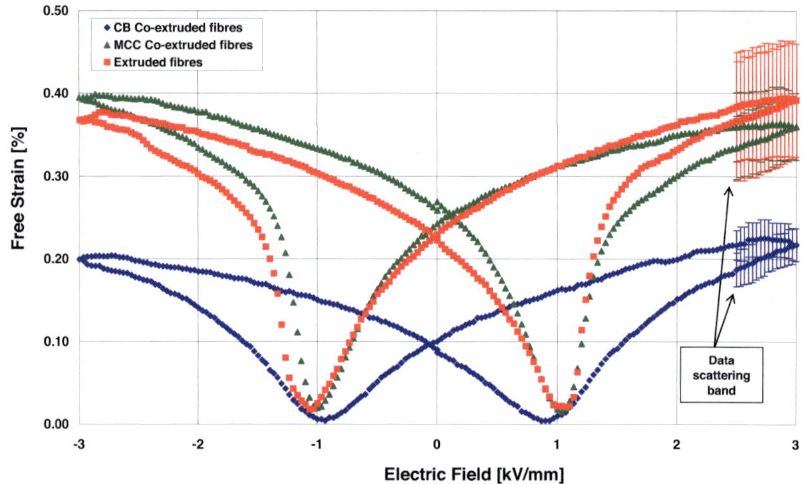

Figure 6.6: Free strain as a function of electric field for extruded and co-extruded PZT fibres. The butterfly curves represent the average values recorded for ten fibres of each process condition.

The MCC co-extruded fibres reached a comparable average maximum free strain as the fibres processed by the conventional extrusion technique (extruded fibres), whereas a drop of ~ 40% in S_{max} was measured for the CB co-extruded fibres. The remanent strain indicates the difference between the unpoled and the poled states at zero field conditions. The coercive field of the three different types of fibres appeared to be similar, but the shape of the butterfly loop for the CB co-extruded fibre looks more like an acceptor doped material with a shallower minimum at the coercive field.

Table 6.2: Average maximum free strain (S_{max}, at 3 kV/mm), remnant strain (S_{rem}), coercive field (E_c) and the effective high field piezoelectric coefficient (d_{33}^*) for the extruded and co-extruded fibres

Fibre	S_{max} [%]	S_{rem} [%]	E_c [kV/mm]	d_{33}^* [pm/V]
Extruded	0.392 ± 0.068	0.226 ± 0.026	1.034 ± 0.021	553 ± 243
CB co-extruded	0.217 ± 0.020	0.097 ± 0.016	0.966 ± 0.001	400 ± 85
MCC co-extruded	0.360 ± 0.040	0.253 ± 0.028	0.966 ± 0.001	357 ± 163

The electromechanical response of PZT materials depends mainly on their porosity, grain size and phase composition. A high porosity, a small grain size and a phase composition away from the morphotropic phase boundary results in a lower electromechanical performance. In view of the microstructure properties (Table 6.1), a slightly higher porosity was observed for the CB co-extruded fibres when compared with the differently processed fibres. However, this might not be the only contribution to their inferior average maximum free strain (Table 6.2). For example, Heiber et al. **[Hei09]** report that similar diameter sintered PZT fibres, with a difference in porosity even higher than the one produced in this work, resulted in a difference of only 10% in S_{max}. Thus, it is reasonable to assume that the difference in electromechanical behaviour observed for the CB co-extruded fibres is due to a change in phase composition.

The effective high voltage piezoelectric coefficient d_{33}^* of the nonlinear piezoelectric loop was determined using equation (6.1):

$$d_{33}^* = \frac{S_{max} - S_{rem}}{E_{max}} \qquad (6.1)$$

where E_{max} is the amplitude of the applied electric field. The strain coefficient d_{33}^* represents the average strain per unit of electric field over the cycle **[Kun07]**. A higher piezoelectric coefficient was calculated for the extruded fibres, and, unexpectedly, the CB co-extruded fibres showed a slightly higher d_{33}^* than the MCC co-extruded fibres (Table 6.2). Besides the huge standard deviation attained for the d_{33}^* calculations, the values for the d_{33}^* determined within this work are in good agreement with previous ones reported in the literature for PZT thin fibres **[Ste10]**. It is reasonable to assume that the inferior piezoelectric coefficients displayed by the co-extruded fibres are related to the negative aspects verified during the binder burnout step (discussed in Chapter 5). The introduction of thermal stresses in those fibres during this step of the processing might have contributed to the degradation of the final properties of the co-extruded fibres.

The residues verified at the surface of the debound CB co-extruded fibres (discussed in Chapter 5) were analysed by means of X-ray fluorescence and X-ray diffraction (Appendix 1). A great amount of S, K, Na and Fe, among other impurities, were detected. The manufacturing process of the used carbon black (named furnace black) employs feedstock oil (such as petroleum) as raw material, which usually contains such residual inorganic materials (ash) **[Seb01]**. As discussed in Chapter 2, PZT crystallizes in the perovskite structure, expressed by the general chemical formula ABO_3. Both A^{2+} and B^{4+} site ions could be substituted with an element of the residue of the carbon black. Dopants with lower charge than that of the replaced ions, known as acceptors (hard-dopants), are compensated by oxygen vacancies. In view of the impurities identified, the diffusion of an acceptor dopant into the inner PZT fibre would be probable in the initial stage of sintering, due to the existence of open pores in the green PZT fibre. For example, the K and Fe detected in the residue of the CB material is assumed to substitute the A (Pb^{2+}) and B (Zr^{4+}/Ti^{4+}) position, respectively, creating oxygen vacancies ($V_O^{\bullet\bullet}$) for charge neutrality. Oxygen vacancies restrict the domain walls motion by introducing space charges and internal fields inside PZT grains and by reducing the unit-cell size **[Mou03]**. The used PZT material is a strontium-potassium-niobium co-doped morphotropic PZT, with a donor excess of 1.5 mol% **[Hel99]**. If this material would take up impurities during firing, it would reduce the effective donor content, explaining the degradation in S_{max} compared to MCC co-extruded and extruded fibres. However, if acceptors are incorporated in the crystal structure, a change in grain size would be expected, as shown in Figure 6.7. On the other hand, Table 6.1 reveals that this is not the case. A possible explanation is that the acceptor dopants do not affect the grain size because they enter the perovskite structure during sintering and not during calcination, as it was the case for the investigations shown in Figure 6.7.

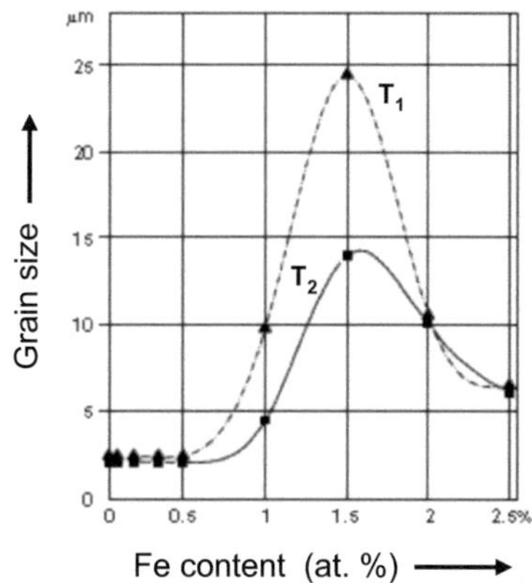

Figure 6.7: Grain size as a function of Fe concentration for the composition $Pb_{1.00}Nd_{0.02}(Zr_{0.545}Ti_{0.455})_{1-y}Fe_yO_3$ for two sintering temperatures (T_1 and T_2) **[Hel08]**.

6.1 Summary

The objective of the work in this chapter was to characterise the sintered PZT solid-fibres obtained by co-extrusion in terms of microstructural properties and electromechanical response in order to compare the data with PZT solid-fibres obtained by conventional extrusion.

Apart from the higher porosity displayed by the co-extruded fibres using carbon black as the filler for the fugitive feedstock (CB co-extruded fibres), the microstructural properties (porosity and grain size) did not change considerably for the different processed fibres. However, a change of the fibre surface phase composition was attained. While the CB co-extruded fibres showed no difference between bulk and fibre surface, the MCC co-extruded and the extruded fibres revealed a higher amount of rhombohedral phase at the surface. The MCC co-extruded fibres reached a comparable average maximum free strain (S_{max} = 0.36%) as the fibres processed by the conventional extrusion technique (S_{max} = 0.39%); however, this value dropped down to 0.22% for the CB co-extruded fibres. The inferior electromechanical performance of the CB co-extruded fibres was related to reduced domain switching, caused by the presence of unfavourable impurities in the carbon black powder. This in turn led to compositionally modified-PZT fibres.

7 Summary

The present work successfully developed a methodology for fabricating lead zirconate titanate thin solid and hollow-fibres by a thermoplastic co-extrusion process. Despite that the current research worked with PZT as the primary material, the method may be transferred to different ceramic materials to manufacture thin fibres and composites. The process steps as well as the relevant aspects determined in the course of these developments are summarized as follows:

a) <u>Selection of the materials to be co-extruded: primary, fugitive and thermoplastic binder system:</u>

Due to the high ratio to surface area that fibres present, with the aim of minimizing lead loss during sintering, the main requirement for the selection of the primary material was to determine a piezoelectric ceramic capable of sintering to full density at a low temperature. In order to fulfil this requirement, two technologically important PZT-based powders (PZT-P505 and PZT-EC65) were characterized regarding their electromechanical performance as a function of sintering temperature. In general, it was verified that the PZT-P505 powder reveals a 10% higher maximum and remnant strain compared with PZT-EC65, independent of the sintering temperature. As PZT-P505 developed a reasonable good microstructure (porosity and grain size) and electromechanical performance even when sintered at 1100 °C, it was selected for this work as primary material for the micro-fabrication of PZT fibres by co-extrusion.

In view of the fugitive material selection, two substances were considered to be investigated within this work: carbon black and microcrystalline cellulose. CB was selected based on the fact that it has been extensively used as the fugitive material in ceramic co-extrusion processes. A suitable CB powder for the present work was selected studying the degradation behaviour of different commercially available powders by means of TGA. On the other hand, MCC was selected as an alternative to carbon black. The selection was done based on the fact that the MCC material experienced complete burn out before sintering of PZT takes place.

Additionally, with the use of stearic acid in the feedstock composition, this substance was able to be mixed with a thermoplastic binder, achieving an apparent viscosity in the range of the primary feedstock (58 vol. % PZT).

Concerning the thermoplastic binder selection, the effects of four different thermoplastic binder systems (polystyrene, low density polyethylene, poly(ethylene-co-ethyl acrylate) and a blend of poly(ethylene-co-ethyl acrylate) and poly(isobutyl methacrylate)) on the processing behaviour of 58 vol. % PZT containing feedstocks were analysed and discussed. The thermal behaviour of the unfilled binders was characterised by means of DSC and TGA. The results attained were essential in determining the optimum temperature range for processing (feedstocks preparation, preform fabrication and co-extrusion). The rheological behaviours of the unfilled polymers in comparison with the corresponding ceramic-filled mixtures were analysed using torque rheometry. Applying an Arrhenius-type approach, the temperature dependency of viscosity was determined and the energy of activation for viscous flow calculated. The Arrhenius representations were shown to be useful in estimating the co-extrusion behaviours of the compounds at a given temperature. The fact that the different thermoplastic binders indicate different interactions with the PZT ceramic resulted in dissimilarities in the activation energies for flow. The die swell effect was investigated on 300 µm diameter PZT fibres obtained by conventional thermoplastic extrusion. Finally, low density polyethylene was selected within this work as the thermoplastic binder material due to its relatively low viscosity with a high solid loading, no indication of degradation, strong particle-polymer interaction as well as comparatively low die swell.

b) <u>Rheological characterisation of the feedstocks:</u>

In order to avoid instabilities during flow, a systematic rheological characterisation of the feedstocks was performed. Applying the Ostwald-de Waele power law model and assuming the torque-rheometer to be analogous to a concentric-cylinder rheometer, the high-shear mixer employed to compound the feedstocks could be additionally used to characterise their rheological behaviour. Empirical viscosity models were used to describe the effect of solids content on the viscosity of highly concentrated suspensions and to estimate the solid loading of a second feedstock at a given viscosity. The following feedstocks, which showed similar rheological behaviour at a given temperature (120 °C) and shear rate (20 rpm) were selected for the co-extrusion process: 58 vol. % PZT as the primary feedstock and 35 vol. % CB or 41 vol. % MCC as the fugitive feedstock.

c) <u>Co-extrusion of the feedstocks:</u>

Successful co-extrusion of solid and hollow-fibres (reduction ratio = 24:1), with well preserved fibre morphologies and defined interfaces between the co-extruded materials were obtained for a viscosity ratio between the feedstocks ranging from 0.98 to 1.16. Outside of this range, interface distortions were observed. In general, co-extrusions carried out at lower shear rates were more satisfactory when compared to high piston speeds, due to the fact that high velocity extrusion may generate unsteady flow conditions.

d) <u>Organic burnout of the co-extruded materials:</u>

The thermal degradation behaviour of the organic vehicles present in the inner and in the outer feedstocks were systematically analysed by using TGA in order to adjust the processing parameters for the debinding step. It could be shown that CB must be decomposed in pure oxygen in order to keep the burnout temperature below 600 °C, whereas MCC decomposes completely in air before sintering of the PZT occurs. However, it is important that the onset temperature for decomposition of the outer layer feedstock is lower than the inner one to avoid the generation of stresses due to the evolution of gaseous species that could not diffuse outwards. In case of solid-fibre preparation these stresses caused fracture of the green fibre, whereas hollow-fibres showed no damage.

e) Sintering and electromechanical behaviour:

The microstructure (porosity, grain size and phase composition) and the electromechanical response of the sintered co-extruded PZT solid-fibres (using both CB and MCC fugitive feedstocks) were compared with PZT solid-fibres of similar diameter (~ 250 µm) obtained by conventional extrusion. Apart from the higher porosity displayed by the co-extruded fibres using CB as the filler for the fugitive feedstock, the porosity and grain size did not change considerably for the different processed fibres. However, a change of the fibre surface phase composition was attained. While the CB co-extruded fibres showed no difference between bulk and fibre surface, the MCC co-extruded and the extruded fibres revealed a higher amount of rhombohedral phase at the surface. The MCC co-extruded fibres reached a comparable average maximum free strain (S_{max} = 0.36%) as the fibres processed by the conventional extrusion technique (S_{max} = 0.39%); however, this value dropped down to 0.22% for the CB co-extruded fibres. This inferior electromechanical performance of the CB co-extruded fibres was related to reduced domain switching, caused by the presence of unfavourable impurities in the CB powder. This in turn led to compositionally modified-PZT fibres.

Summarizing, it was concluded that successful defect-free thermoplastic co-extrusion of ceramics is only achievable if the materials selected, rheology, co-extrusion parameters, debinding and sintering steps are all well defined.

7.1 Recommendations for future studies

Based on the present work, some topics are identified for further research work to contribute to a better understanding of the co-extrusion process of piezoelectric ceramic fibres:

- Study of the influence of the reduction ratio on the processability of co-extruded fibres. Use of different inner rods diameters – larger (carried out in this study), smaller and with the same size as the reduction die.

- Study of the optimisation of the outer feedstock burn-out process through the addition of organic materials which experiences lower temperatures degradation as the selected fugitive filler and/or high molecular weight thermoplastic binder.

- Study of the chemical composition across the sintered co-extruded fibre radius by electron probe micro analysis or other suitable methods.

Moreover, this work has drawn attention to several scientific issues which merit future research. The correlation of the activation energy of viscous flow with the polymer-particle interactions discussed in Chapter 4 could be further addressed working with thermoplastic binders with the same functional group, albeit differing in molecular weight. Additionally, working with amorphous and semi-crystalline thermoplastic binders with similar molecular weight would allow a greater understanding of the influence of the binder crystallinity on the processability of the feedstocks. The use of further analytical methods (e.g., gel permeation chromatography and infrared adsorption spectroscopy) for a specific investigation of the degradation phenomenon of the feedstocks during mixing and extrusion would be of benefit in improving these processing steps as well as in understanding the flow behaviour of the feedstocks. A closer investigation of the viscoelastic behaviour of the feedstocks, performed with a systematic rheological approach using oscillatory tests (rotational rheometers), would be of advantage in developing complex shapes co-extruded materials.

Appendix: Carbon black residue's characterisation

The present section displays the results of the characterizations ((a) X-ray diffraction analyses and (b) X-ray fluorescence analyses) performed on the residues left on the surface of the debound CB co-extruded fibres (discussed in Chapter 5), showing the impurities detected. The X-ray diffraction was carried out using a PANanalytical XPert Pro diffractometer from Phillips, equipped with a Cu Kα radiation source (as described in Chapter 3). The wavelength-dispersive X-ray fluorescence instrument used (Philips PW2400) is a sequential spectrometer with an end-window Rh X-ray tube, 3000W maximum power, 60 kV maximum voltage and 125 mA maximum current.

(a) X-ray diffraction analyses

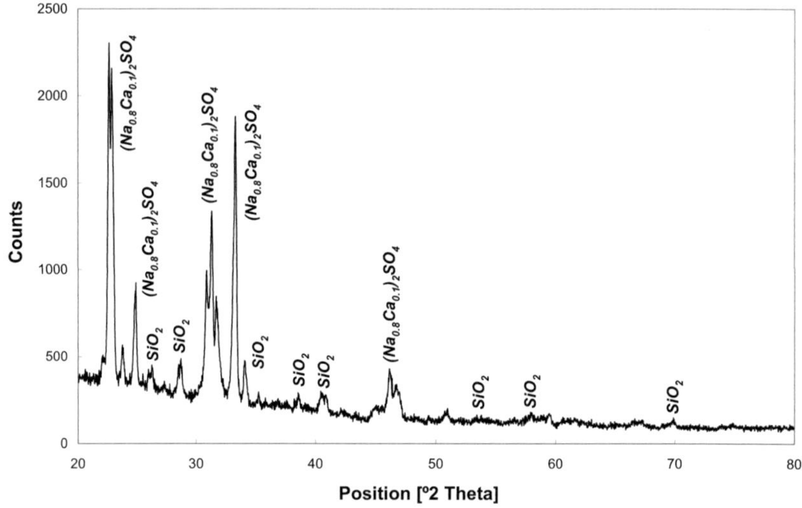

(b) X-ray fluorescence analyses

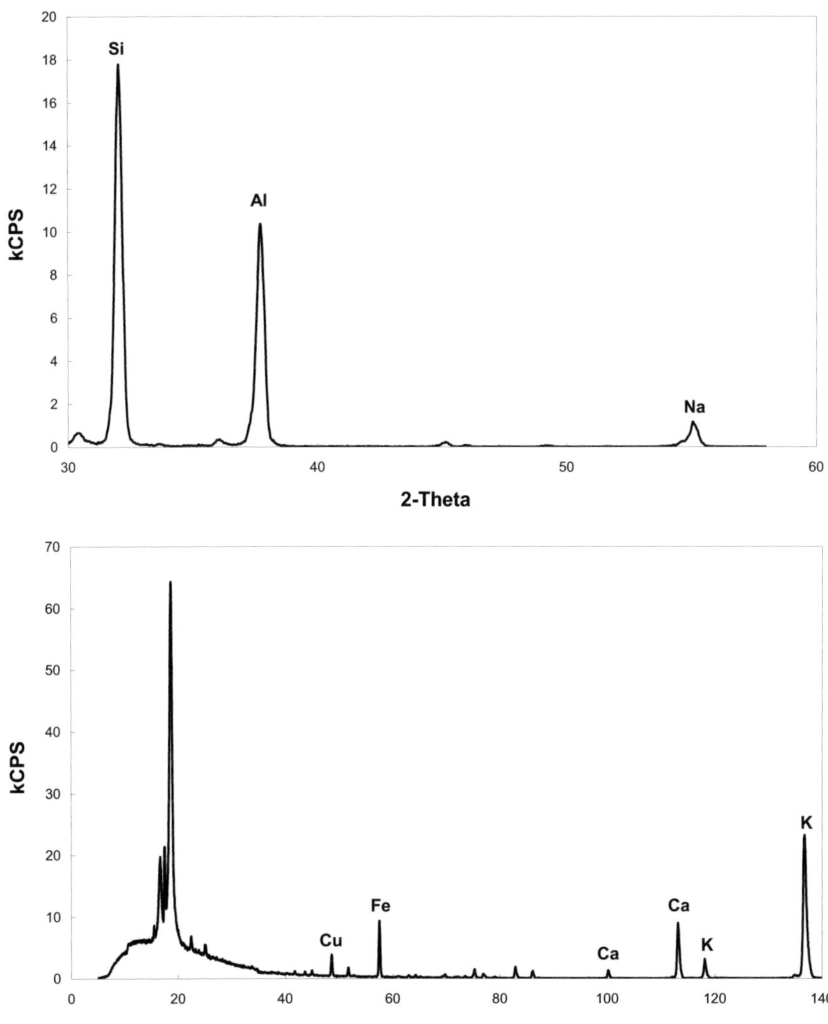

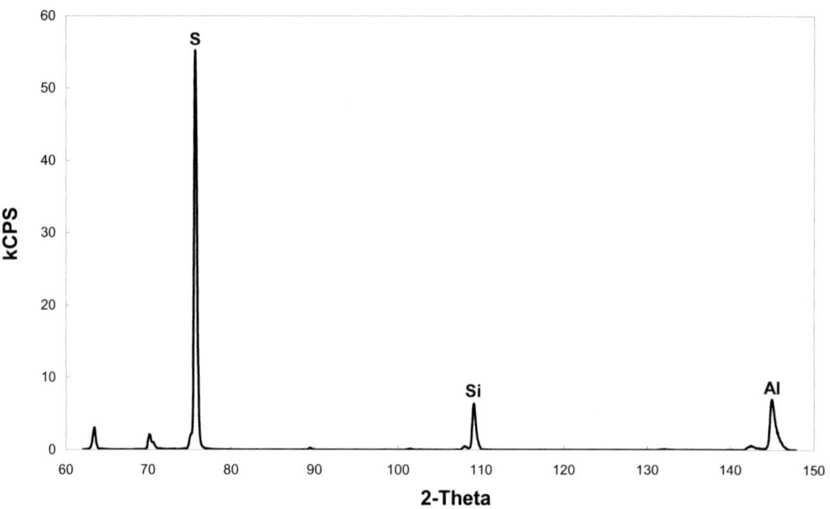

kCPS
60
50
40
30
20
10
0
S
Si
Al
60
70
80
90
100
110
120
130
140
150
2-Theta

References

[Akd05] Akdogan, E. K. / Allahverdi, M. / Safari, A. (2005)
Piezoelectric composites for sensor and actuator applications
IEEE Transactions on Ultrasonics, Ferroelectrics and Frequency Control
Vol. 52, p. 746-775.

[Akd08] Akdogan, E. K. and Safari, A. (2008)
Thermodynamics of ferroelectricity
in Piezoelectric and Acoustic Materials for Transducer Applications, Safari, A.
and Akdogan, E. K. (eds)
Springer Science + Business Media, LLC.

[Aul89] Auld, B. (1989)
Waves and vibrations in periodic piezoelectric composite materials
Materials Science and Engineering: A
Vol. 122, p. 65-70

[Ban97] Bandyopadhyay, A. / Panda, R. K. / Janas, V. F. / Agarwala, M. K. /
Danforth, S. C. / Safari, A. (1997)
Processing of piezocomposites by fused deposition technique
Journal of the American Ceramic Society
Vol. 80, p. 1366-1372.

[Bat75] Battista, O. A. (1975)
Microcrsytal polymer science
Copyright © by McGraw-Hill, Inc.,United States of America.

[Bee02] Beeaff, D. R. and Hilmas, G. E. (2002)
Rheological behaviour of coextruded multilayer archictures
Journal of Material Science
Vol. 37, p. 1259-1264.

[Bel09] Belloli, A. / Heiber, J. / Clemens, F. / Ermanni, P. (2009)
Novel characterization procedure for single piezoelectric fibers
Journal of Intelligent Material Systems and Structures
Vol. 20, p. 355-363.

[Ben93] Benbow, J. and Bridgwater, J. (1993)
Paste flow and extrusion
Oxford Series of Advanced Manufacturing, Clarendon Press, Oxford.

[Ben97] Bent, A. A. and Hagood, N. W. (1997)
Piezoelectric fibre composites with interdigitated electrodes
Journal of Intelligent Material Systems and Structures
Vol. 8, p. 903-919.

[Ben00] Bent, A. A. and Pizzochero, A. E. (2000)
Recent advances in active fiber composites for structural control
Proceedings of SPIE
Vol. 3991, p. 244-254.

[Big83] Bigg, D. M. (1983)
Rheological behaviour of highly filled polymer melts
Polymer Engineering and Science
Vol. 23, p. 206-210.

[Bla08] Blackburn, S. and Wilson, D. I. (2008)

Shaping ceramics by plastic processing

Journal of the European Ceramic Society

Vol. 28, p. 1341-1351.

[Bly67] Blyler Jr., L. L. and Daane, J. H. (1967)

An analysis of brabander torque rheometer data

Polymer Engineering and Science

Vol. 7, p. 178-181.

[Bre04] Brei, D. and Cannon, B. J. (2004)

Piezoeceramic hollow fiber active composites

Composites Science and Technology

Vol. 64, p. 245-261

[Bru05] Brunner, A. J. / Barbezat, M. / Huber, C. / Flüeler, P. H. (2005)

The potential of active fiber composites for actuating and sensing applications

in structural health monitoring

Materials and Structures

Vol. 38, p. 561-567.

[Bry99] Brydson, J. A. (1999)

Plastic materials

Butterworth-Heinemann, Oxford, Seventh Edition.

[Buc01] Buchtel, A. M. and Earl, D. A. (2001)

Cordierite honeycomb structures formed by reduction extrusion processing

Journal of Materials Science Letters

Vol. 20, p. 1759-1761.

[Che92] Cheng, S. / Lloyd, I. K. / Kahn, M. (1992)

Modification of surface texture by grinding and polishing lead zirconate titanate ceramics

Journal of the American Ceramic Society

Vol. 75, p. 2293-2296.

[Che01] Chen, Z. / Ikeda, K. / Murakami, T. / Takeda, T. (2001)

Extrusion behavior of metal-ceramic composite pipes in multi-billet extrusion process

Journal of Materials Processing Technology

Vol. 114, p. 154-60.

[Che08] Chen, Y. / Struble, L. J. / Paulino, H. (2008)

Using rheology to achieve co-extrusion of cement-based materials with graded cellular structures.

International Journal of Applied Ceramic Technology

Vol. 5, p. 513-521.

[Chi02] Chinn, R. E. (2002)

Ceramography: Preparation and analysis of ceramic microstructures

ASM International®, Materials Park, OH 44073-0002.

[Cho71] Chong, J. S. / Christiansen, E. B. / Baer, A. D. (1971)

Rheology of concentrated suspensions

Journal of Applied Polymer Science

Vol. 15, p. 2007-2021.

[Cle07a] Clemens, F. (2007)

Thermoplastic extrusion for ceramic bodies

in Extrusion in Ceramics, Händle, F. (Ed.)

Springer-Verlag Berlin Heidelberg.

[Cle07b] Clemens, F. J. / Wallquist, V. / Buchser, W. / Wegmann, M. / Graule, T. (2007)
Silicon carbide fiber-shaped microtools by extrusion and sintering SiC with and without carbon powder sintering additive
Ceramics International
Vol. 33, p. 491-496.

[Cog81] Cogswell, F. N. (1981)
Polymer melt rheology
John Wiley & Sons, New York & Toronto.

[Cru98] Crumm, A. T. and Halloran, J. W. (1998)
Fabrication of microconfigured multicomponent ceramics
Journal of the American Ceramic Society
Vol. 81, p. 1053-1057.

[Cru07] Crumm, A. T. / Halloran, J. W. / Silva, E. C. N. / Espinosa, F. M. (2007)
Microconfigured piezoelectric artificial materials for hydrophones
Journal of Materials Science
Vol. 42, p. 3944-3950.

[Den05] Dent, A. C. / Nelson, L. J. /Bowen, C. R. / Stevens, R. / Cain, M. / Stewart, M. (2005)
Characterisation and properties of fine scale PZT fibres
Journal of the European Ceramic Society
Vol. 25, p. 2387-2391.

[Dit10] Dittmer, R. / Clemens, F. / Schönecker, A. / Scheithauer. U. / Ismael, M. R. / Graule, T. (2010)

Microstructure analysis and mechanical properties of $Pb(Zr,Ti)O_3$ fibers derived from different processing routes

Journal of the American Ceramic Society

Vol. 93, p. 2403-2410.

[Doo02] Dooley, J. (2002)

Viscoelastic flow effects in multilayer polymer coextrusion

PhD Thesis, Eindhoven University of Technology.

[Doo03] Dooley, J. and Rudolph, L. (2003)

Viscous and elastic effects in polymer coextrusion

Journal of Plastic Film and Sheeting

Vol. 19, p. 111-122.

[Don04] Donev, A. et al. (2004)

Improving the density of jammed disordered packings using ellipsoids

Science

Vol. 303, p. 990-993.

[Dro09] Droushiotis, N. / Othman, M. H. D. / Doraswami, U. / Wu, Z. / Kelsall, G. / Li, K. (2009)

Novel co-extruded electrolyte-anode hollow fibres for solid oxide fuel cells

Electrochemistry Communications

Vol. 11, p. 1799-1802.

[Edi86] Edirisinghe, M. J. and Evans, J. R. G. (1986)
Review: Fabrication of engineering ceramics by injection molding. I. Materials selection
International Journal of High Technology Ceramics
Vol. 2, p. 1-31.

[Fel04] Feller, J. F. / Langevin, D. / Marais, S. (2004)
Influence of processing conditions on sensitivity of conductive polymer composites to organic solvent vapours
Synthetic Metals
Vol. 144, p. 81-88.

[Fer95] Fernandes, J. C. / Hall, D. A. / Cockburn, M. R. / Greaves, G. N. (1995)
Phase coexistence in PZT ceramic powders
Nuclear Instruments and Methods in Physics Research B
Vol. 97, p. 137-141.

[Fer96] Fernandez, J. F. / Dogan, A. / Zhang, Q. M. / Tressler, J. F. / Newnham, R. E. (1996)
Hollow piezoelectric composites
Sensors and Actuators A
Vol. 51, p. 183 – 192.

[Fra67] Frankel, N. A. and Acrivos, A. (1967)
On the viscosity of a concentrated suspension of solid spheres
Chemical Engineering Science
Vol. 22, p. 847-853.

[Fre98] French, J. D. and Cass, R. B. (1998)
Developing innovative ceramic fibers
American Ceramic Society Bulletin
Vol. 76, p. 61-65.

[Gen94] Gentilman, R. L. / Fiore, D. F. / Pham, H. T. / French, K. W. / Bowen, L. J. (1994)
Fabrication and properties of 1-3 PZT-polymer composites
in Ceramic Transactions, Ferroic Materials: Design, Preparation, and Characteristics.
American Ceramic Society, Westerville, OH.
Vol. 43, p. 239-247.

[Ger96] German, R. M. (1996)
Sintering theory and practice
New York: John Wiley & Sons, INC.

[Goo67] Goodrich, E. and Porter, R. S. (1967)
A rheological interpretation of torque-rheometer data
Polymer Engineering and Science
Vol. 7, p. 45-51.

[Gre01] Greenwood, R. / Kendall, K. / Bellon, O. (2001)
A method for making alumina fibres by co-extrusion of an alumina and starch paste
Journal of the European Ceramic Society
Vol. 21, p. 507-513.

[Gur94] Gururaja, T. R. (1994)

Piezoelectrics for medical ultrasonic imaging

American Ceramic Society Bulletin

Vol. 73, p. 50-55.

[Hal04] D. A. Hall, A. Steuwer, B. Cherdhirunkorn, T. Mori, and P. J. Withers (2004)

A high energy synchrotron x-ray study of crystallographic texture and lattice strain in soft lead zirconate titanate ceramics

Journal of applied physics

Vol. 96, p. 4245-4252

[Ham98] Hammer, M. and Hoffmann, M. J. (1998)

Sintering model for mixed-oxide derived lead zirconate titanate ceramics

Journal of the American Ceramic Society

Vol. 81, p. 3277-3284.

[Han04] Hansch, R. / Seifert, S. / Braue, W. / Sporn, D. / Müller, G. (2004)

The effects of the PbO content upon the microstructure and the ferroelectric properties of undoped sol-gel derived PZT (53/47) fibers

Journal of the European Ceramic Society

Vol. 24, p. 2485-2497.

[Hea97] Hearn, E. J. (1997)

Mechanics of materials I - An introduction to the mechanics of elastic and plastic deformation of solids and structural materials

Butterworth-Heinemann, Oxford, Third edition.

[Hei05] Heiber, J. / Clemens, F. / Graule, T. / Hülsenberg, D. (2005)
Thermoplastic extrusion to highly-loaded thin green fibres containing $Pb(Zr,Ti)O_3$
Advanced Engineering Materials
Vol. 7, p. 404-408.

[Hei07] Heiber, J. / Clemens, F. / Helbig, U. / de Meuron, A. / Soltmann, C. / Graule, T. / Hülsenberg, D. (2007)
Properties of $Pb(Zr, Ti)O_3$ fibres with a radial gradient structure
Acta Materialia
Vol. 55, p. 6499-6506.

[Hei08] Heiber, J. / Clemens, F. / Graule, T. / Hülsenberg, D. (2008)
Influence of varying the powder loading content on the homogeneity and properties of extruded PZT-fibers
Key Engineering Materials
Vol. 368-378, p. 11-14, Part 1-2.

[Hei09] Heiber, J. / Belloli, A. / Ermanni, P. / Clemens, F. (2009)
Ferroelectric characterisation of single PZT fibers
Journal of Intelligent Materials Systems and Structures
Vol. 20, p. 379-385.

[Hel99] Helke, G. / Seifert, S. / Cho, S. J. (1999)
Phenomenological and structural properties of piezoelectric ceramics based on $xPb(Zr, Ti)O_3-(1-x)Sr(K_{0.25}Nb_{0.75})O_3$ (PZT/SKN) solid solutions
Journal of the European Ceramic Society
Vol. 19, p. 1265-1268.

[Hel08] Helke, G. and Lubitz, K. (2008)
Piezoelectric PZT ceramics
in Piezoelectricity – Evolution and future of a technology, Heywang, W., Lubitz, K. and Wersing, W. (eds)
Springer Series in MATERIALS SCIENCE

[Hil05] Hilmas, G.E. / Platero, M. / Popovich, D. / Rigali, M.J. (2005)
Ceramic components having multilayered architectures and processes for manufacturing the same
United States Patent Application Publication
Publication number: US 2005/0082726 A1

[Hof01] Hoffmann, M. / Hammer, M. / Endriss, A. / Lupascu, D. C. (2001)
Correlation between microstructure, strain behavior, and acoustic emission of soft PZT ceramics
Acta Materialia
Vol. 49, p. 1301-1310.

[Hrd98] Hrdina, K. E. and Halloran, J. W. (1998)
Dimensional changes during binder removal in a mouldable ceramic system
Journal of Materials Science
Vol. 33, p. 2805-2815.

[Jaf71] Jaffe, B. / Cook, W. R. / Jaffe, H. (1971)
Piezoelectric ceramics
Academic Press London & New York.

[Jan95a] Janas, V. F. / McNulty, T. F. / Walker, F. R. / Schaeffer, R. P. / Safari, A. (1995)
Processing of 1-3 piezoelectric ceramic/polymer composites
Journal of the American Ceramic Society
Vol. 78, p. 2425-2430.

[Jan95b] Janas, V. F. and Safari, A. (1995)
Overview of fine-scale piezoelectric ceramic/polymer composite processing
Journal of the American Ceramic Society
Vol. 78, p. 2945-2955.

[Kah92] Kahn, M. and Chase, M. (1992)
Effects of heat treatments on multilayer piezoelectric ceramic-air composites
Journal of the American Ceramic Society
Vol. 75, p. 649-656.

[Kim04] Kim, D. K. and Kriven, W. M. (2004)
Mullite ($3Al_2O_3.2SiO_2$)-aluminum phospahte ($AlPO_4$), oxide, fibrous monolithic composites
Journal of the American Ceramic Society
Vol. 87, p. 794-803.

[Kin83a] Kingon, A. I. and Clark, J. B. (1983)
Sintering of PZT ceramics: I, Atmosphere control
Journal of the American Ceramic Society
Vol. 66, p. 253-256.

[Kin83b] Kingon, A. I. and Clark, J. B. (1983)
Sintering of PZT ceramics: II, Effect of PbO content on densification kinetics
Journal of the American Ceramic Society
Vol. 66, p. 256-260.

[Kna06] Knapp, A. M. and Halloran, J. W. (2006)
Binder removal from ceramic-filled thermoplastic blends
Journal of the American Ceramic Society
Vol. 89, p. 2776-2781.

[Koh02] Koh, Y. H. / Kim, H. W. / Kim, H. E. / Halloran, J. W. (2004)
Fabrication of macrochannelled-hydrohyapatite bioceramic by a coextrusion process
Journal of the American Ceramic Society
Vol. 85, p. 2578-2580.

[Koh04] Koh, Y. H. and Halloran, J. W. (2004)
Green machining of a thermoplastic ceramic-ethylene ethyl acrylate/isobutyl methacrylate compound
Journal of the American Ceramic Society
Vol. 87, p. 1575-1577.

[Kor04] Kornmann, X. and Huber, C. (2004)
Microstructure and mechanical properties of PZT fibres
Journal of the European Ceramic Society
Vol. 24, p. 1987-1991.

[Kov97] Kovar, D. / King, B. H. / Trice, R. W. / Halloran, J. W. (1997)
Fibrous monolithic ceramics
Journal of the American Ceramic Society
Vol. 80, p. 2471-2487.

[Kri59] Krieger, I. M. and Dougherty, T. J. (1959)
A mechanism for non-Newtonian flow in suspensions of rigid spheres
Transaction of the Society of Rheology
Vol. 3, p. 137-152.

[Kun05] Kungl, H. (2005)

Dehnungsverhalten von morphotropem PZT

PhD Thesis, Universität Karlsruhe.

[Kun07] Kungl, H. / Fett, T. / Wagner, S. / Hoffmann, M. (2007)

Nonlinearity of strain and strain hysteresis in morphotropic LaSr-doped lead zirconate titanate under unipolar cycling with high electric fields

Journal of Applied Physics

Vol. 101, 044101

[Kun07] Kungl, H. and Hoffmann, M. (2010)

Effects of sintering temperature on microstructure and high field strain of niobium-strontium doped morphotropic lead zirconate titanate

Journal of Applied Physics

Vol. 107, 054111

[Lam09] Lamnawar, K. and Maazouz, A. (2009)

Role of the interphase in the flow stability of reactive coextruded multilayer polymers

Polymer Engineering and Science

Vol. 49, p. 727-739.

[Let08] Lethiecq, M. / Levassort, F. / Certon, D. / Hue, L. P. T. H. (2008)

Piezoelectric transducer desing for medical diagnosis and NDE

in Piezoelectric and Acoustic Materials for Transducer Applications, Safari, A. and Akdogan, E. K. (eds)

Springer Science + Business Media, LLC.

[Lev89] Levy, S. and Carley, J. F. (1989)

Plastics extrusion technology handbook

Industrial Press, New York, Second Edition.

[Lew97] Lewis, J. A. (1997)
Binder removal from ceramics
Annual Review of Materials Science
Vol. 27, p. 147-173.

[Lew00] Lewis, J. A. (2000)
Colloidal processing of ceramics
Journal of the American Ceramic Society
Vol. 83, p. 2341-2359

[Lia01] Liang, Z. and Blackburn, S. (2001)
Design and characterization of a co-extruder to produce tri-layer ceramic tubes
semi-continuously
Journal of the European Ceramic Society
Vol. 21, p. 883-892.

[Mac94] Macosko, C. W. (1994)
Rheology: principles, measurements, and applications
Wiley-VCH, Canada.

[Mck08] McKinstry, T. S. (2008)
Crystal chemistry of piezoelectric materials
in Piezoelectric and Acoustic Materials for Transducer Applications, Safari, A.
and Akdogan, E. K. (eds)
Springer Science + Business Media, LLC.

[Mcn99] McNulty, T. F. / Shanefield, D. J. / Danforth, S. C. / Safari, A. (1999)
Dispersion of Lead zirconate titanate for fused deposition of ceramics
Journal of the American Ceramic Society
Vol. 82, p. 1757-1760.

[Mey98] Meyer, R. J. / Shrout, T. R. / Yoshikawa, S. (1998)

Lead zirconate titanate fine fibers derived from alkoxide-based sol-gel technology

Journal of the American Ceramic Society

Vol. 81, p. 861-868.

[Min75] Minagawa, N. and White, J. L. (1975)

Co-extrusion of unfilled and TiO_2-filled polyethylene: influence of viscosity and die cross-section on interface shape

Polymer Engineering and Science

Vol. 15, p. 825-830.

[Moo72] Moore, W. J. (1972)

Physical chemistry

Orient Longman Limited, New Jersey, Fifth Edition.

[Mou03] Moulson, M. J. and Herbet, J. M. (2003)

Electroceramics – materials, properties, applications

John Wiley & Sons, England, Second Edition.

[Nel02] Nelson, L. J. (2002)

Smart piezoelectric fibre composites

Materials Science and Technology

Vol. 18, p. 1245-1256.

[New78] Newnham, R. E. / Skinner, D. P. / Cross, L. E. (1978)

Connectivity and piezoelectric-pyroelectric composites

Materials Research Bulletin

Vol. 13, p. 525-536.

[Oka00] Okada, K. and Nagase, Y. (2000)
Viscosity and powder dispersion in ceramic injection molding mixtures
Journal of Chemical Engineering of Japan
Vol. 33, p. 168-173.

[Pal07] Palaniyandi, P. and Simonsen, J. (2007)
Effect of compatibilizers on the crystallization kinetics of cellulose-filled high density polyethylene
Composite Interfaces
Vol. 14, p. 73-83.

[Pay98] Payne, C. E. A. and Evans, J. R. G. (1998)
Cross-flow ceramic structures produced by plastic processing
International Journal of Advanced Manufacturing Technology
Vol. 14, p. 508-513.

[Que77] Quemada, D. (1977)
Rheology of concentrated disperse systems and minimum energy dissipation principle. I. Viscosity-concentration relationship
Rheologica Acta
Vol.16, p. 82-94.

[Ree95] Reed, J. S. (1995)
Principles of ceramics processing
John Wiley & Sons, Inc., New York, Second Edition.

[Sav81] Savakus, H. P. / Klicker, K. A. / Newnham, R. E. (1981)
PZT-Epoxy piezoelectric transducers: A simplified fabrication procedure
Materials Research Bulletin
Vol. 16, p. 677-680.

[Sch78] Schrenk, W. J. / Bradley, N. L. / Alfrey Jr, T. / Maack, H. (1978)

Interfacial flow instability in multilayer coextrusion

Polymer Engineering and Science

Vol. **18**, p. 620-623.

[Sch04] Schramm, G. (2004)

A practical approach to rheology and rheometry

Thermo Electron (Karlsruhe) GmbH, Germany, Second Edition.

[Sch08] Schönecker, A. (2008)

Piezoelectric fiber composite fabrication

in Piezoelectric and Acoustic Materials for Transducer Applications, Safari, A.
and Akdogan, E. K. (eds)

Springer Science + Business Media, LLC.

[Schö08] Schönau, K. A. (2008)

In situ Synchrotron diffraction of lead-zirconate-titanate at its morphotropic
phase boundary

PhD Thesis, Technischem Universität Darmstadt

[Seb01] Sebok, E. B. and Taylor, R. L. (2001)

Carbon blacks

in Encyclopaedia of Materials: Science and Technology

p. 902 – 906.

[She99] Shenoy, A. V. (1999)

Rheology of filled polymer systems

Academic Publishers, Netherlands.

[Sim03] Simon, G. P. (2003)

Polymer characterization techniques and their application to blends

American Chemical Society, Washington, D.C., Oxford University Press.

[Ste10] Steinhausen, R. / Kerna, S. / Pientschke, C. / Beige, H. / Clemens, F. / Heiber, J. (2010)

A new measurement method of piezoelectric properties of single ceramic fibres

Journal of the European Ceramic Society

Vol. 30, p. 205-209.

[Str99] Strock, H. B. / Pascucci, M. R. / Parish, M. V. / Bent, A. A. / Shrout, T. R. (1999)

Active PZT fibers, a commercial production process

SPIE Conference Proceeding

Vol. 3675, p. 22-31.

[Täf04] Täffner, U. / Carle, V. / Scäfer, U. / Hoffmann, M. J. (2004)

Preparation and microstructural analysis of high-performance ceramics

ASM Handbook Volume 9: Metallography and Microstructures.

[Tak98] Takase, M. / Kihara, S. I. / Funatsu, K. (1998)

Three-dimensional viscoelastic numerical analysis of the encapsulation phenomena in coextrusion

Rheologica Acta

Vol. 37, p. 624-634.

[Tul02] Tuller, H. and Avrahami, Y. (2002)

Electroceramics

in Encyclopedia of Smart Materials

John Wiley & Sons, Inc. New York.

[Van98] Van Hoy, C. / Barda, A. / Griffith, M. / Halloran, J. W. (1998)
Microfabrication of ceramics by co-extrusion
Journal of the American Ceramic Society
Vol. 81, p.152-158.

[Wal92] Waller, D. J. and Safari, A. (1992)
Piezolectric lead zirconate titanate ceramic fiber/polymer composites
Journal of the American Ceramic Society
Vol. 75, p.1648-1655.

[Weg98] Wegmann, M. / Gut, B. / Berroth, K. (1998)
Extrusion of polycrystalline ceramic fibers
CFI - Ceramic Forum International
Vol. 75, p. 35-37.

[Wig01] Wight Jr., J. F. and Reed, J. S. (2001)
Polymer-plasticized ceramic extrusion, Part 1
American Ceramic Society Bulletin
Vol. 80, p. 31-35.

[Yoo05] Yoon, C. B. / Koh, Y. H. / Park, G. T. / Kim, H. E. (2005)
Multilayer actuator composed of PZN-PZT and PZN-PZT/Ag fabricated by co-extrusion process
Journal of the American Ceramic Society
Vol. 88, p.1625-1627.

[Zat05] Zatloukal, M. / Kopytko, W. / Lengálová, A. / Vlcek, J. (2005)
Theoretical and experimental analysis on interfacial instabilities in coextrusion flows
Journal of Applied Polymer Science
Vol. 98, p. 153-162.